Lecture Notes of
the Unione Matematica Italiana

9

Maria Evelina Rossi • Giuseppe Valla

Hilbert Functions
of Filtered Modules

 Springer

Prof. Maria Evelina Rossi
University of Genoa
Department of Mathematics
Via Dodecaneso 35
16146 Genova
Italy
rossim@dima.unige.it

Prof. Giuseppe Valla
University of Genoa
Department of Mathematics
Via Dodecaneso 35
16146 Genova
Italy
valla@dima.unige.it

ISSN 1862-9113
ISBN 978-3-642-14239-0   e-ISBN 978-3-642-14240-6
DOI 10.1007/978-3-642-14240-6
Springer Heidelberg Dordrecht London New York

Library of Congress Control Number: 2010933510

Mathematics Subject Classification (2000): 13A02, 13A30, 13C14, 13C15, 13H15, 14B99

Cover design: SPi Publisher Services

Printed on acid-free paper

Springer is part of Springer Science+Business Media (www.springer.com)

Preface

Hilbert Functions play major roles in Algebraic Geometry and Commutative Algebra, and are becoming increasingly important also in Computational Algebra. They capture many useful numerical characters associated to a projective variety or to a filtered module over a local ring.

Starting from the pioneering work of D.G. Northcott and J. Sally, we aim to gather together in one place many new developments of this theory by using a unifying approach which gives self-contained and easier proofs.

The extension of the theory to the case of general filtrations on a module, and its application to the study of certain graded algebras which are not associated to a filtration are two of the main features of the monograph.

The material is intended for graduate students and researchers who are interested in Commutative Algebra, in particular in the theory of the Hilbert functions and related topics.

Genoa,   *Maria Evelina Rossi*
March, 2010   *Giuseppe Valla*

Acknowledgements

We would like to take this opportunity to thank sincerely Judith Sally because her work has had such a strong influence on our research into these subjects. In particular, several problems, techniques and ideas presented in this text came from a careful reading of her papers, which are always rich in examples and motivating applications.

Let us also not forget the many other colleagues who over the years have shared their ideas on these topics with us. Some of them were directly involved as co-authors in joint research reported here, while others gave a substantial contribution via their publications and discussions.

Contents

Introduction

The notion of Hilbert function is central in commutative algebra and is becoming increasingly important in algebraic geometry and in computational algebra. In this presentation we shall deal with some aspects of the theory of Hilbert functions of modules over local rings, and we intend to guide the reader along one of the possible routes through the last three decades of progress in this area of dynamic mathematical activity.

Motivated by the ever increasing interest in this field, our aim is to gather together many new developments of this theory in one place, and to present them using a unifying approach which gives self-contained and easier proofs. In this text we shall discuss many results by different authors, following essentially the direction typified by the pioneering work of J. Sally (see [86–93]). Our personal view of the subject is most visibly expressed by the presentation of Chaps. 1 and 2 in which we discuss the use of the superficial elements and related devices.

Basic techniques will be stressed with the aim of reproving recent results by using a more elementary approach. This choice was made at the expense of certain results and various interesting aspects of the topic that, in this presentation, must remain peripheral. We apologize to those whose work we may have failed to cite properly.

The material is intended for graduate students and researchers who are interested in Commutative Algebra, in particular in results on the Hilbert function and the Hilbert polynomial of a local ring, and applications of these. The aim was not to write a book on the subject, but rather to collect results and problems inspired by specialized lecture courses and schools recently delivered by the authors. We hope the reader will appreciate the large number of examples and the rich bibliography.

Starting from classical results of D. Northcott, P. Samuel, S. Abhyankar, E. Matlis and J. Sally, many papers have been written on this topic which is considered an important part of the theory of blowing-up rings. This is because the Hilbert function of the local ring $(A, \mathfrak{m})$ is by definition the numerical function $H_A(t) := \dim_k(\mathfrak{m}^t/\mathfrak{m}^{t+1})$, hence it coincides with the classical Hilbert function of the standard graded algebra $gr_\mathfrak{m}(A) := \oplus_{t \geq 0} \mathfrak{m}^t/\mathfrak{m}^{t+1}$, the so-called *tangent cone* of A for the reason that we shall explain later. The problems arise because, in passing from

A to $gr_m(A)$, we may lose many good properties, such as being a complete intersection, being Cohen–Macaulay or Gorenstein.

Despite the fact that the Hilbert function of a standard graded algebra A is well understood when A is Cohen–Macaulay, very little is known when it is a local Cohen–Macaulay ring. One of the main problems is whether geometric and homological properties of the local ring A can be carried on the corresponding tangent cone $gr_m(A)$. For example if a given local domain has fairly good properties, such as normality or Cohen–Macaulayness, its depth provides in general no information on the depth of the associated graded ring. It could be interesting to remind that an open problem is to characterize the Hilbert function of an affine curve in $\mathbf{A}^3$ whose defining ideal is a complete intersection, while a well known formula gives the Hilbert function of any complete intersection of homogeneous forms in terms of their degrees.

The Hilbert function of a local ring (A, m) is a classical invariant which gives information on the corresponding singularity. The reason is that the graded algebra $gr_m(A)$ corresponds to an important geometric construction: namely, if A is the localization at the origin of the coordinate ring of an affine variety V passing through 0, then $gr_m(A)$ is the coordinate ring of the tangent cone of V, that is the cone composed of all lines that are limiting positions of secant lines to V in 0. The *Proj* of this algebra can also be seen as the exceptional set of the blowing-up of V in 0.

Other graded algebras come into the picture for different reasons, for example the Rees algebra, the Symmetric algebra, the Sally module and the Fiber Cone. All these algebras are doubly interesting because on one side they have a deep geometrical meaning, on the other side they are employed for detecting basic numerical characters of the ideals in the local ring (A, m). Therefore, much attention has been paid in the past to determining under which circumstances these objects have a good structure.

In some cases the natural extension of these results to m-primary ideals has been achieved, starting from the fundamental work of P. Samuel on multiplicities. More recently the generalization to the case of a descending multiplicative filtration of ideals of the local ring A has now become of crucial importance. For example, the Ratliff–Rush filtration (cf. papers by J. Elias, W. Heinzer, S. Huchaba, S. Itoh, T. Marley, T. Puthenpurakal, L.J. Ratliff–D. Rush, M.E. Rossi, J. Sally, G. Valla) and the filtration given by the integral closure of the powers of an ideal (cf. papers by A. Corso, S. Itoh, C. Huneke, C. Polini, B. Ulrich, W. Vasconcelos, J. Verma) are fundamental tools in much of the recent work on blowing-up rings.

Even though of intrinsic interest, the extension to modules has been largely overlooked, probably because, even in the classical case, many problems were already so difficult. Nevertheless, a number of results have been obtained in this direction: some of the work done by D. Northcott, J. Fillmore, C. Rhodes, D. Kirby, H. Meheran and, more recently, T. Cortadellas and S. Zarzuela, A.V. Jayanthan and J. Verma, T. Puthenpurakal, V. Trivedi has been carried over to the general setting.

We remark that the graded algebra $gr_m(A)$ can also be seen as the graded algebra associated to an ideal filtration of the ring itself, namely the m-adic filtration

$\{\mathfrak{m}^j\}_{j\geq 0}$. This gives an indication of a possible natural extension of the theory to general filtrations of a finite module over the local ring $(A, \mathfrak{m})$.

Let A be a commutative noetherian local ring with maximal ideal $\mathfrak{m}$ and let M be a finitely generated A-module. Let $\mathfrak{q}$ be an ideal of A; a q-filtration $\mathbb{M}$ of M is a collection of submodules M_j such that

$$M = M_0 \supseteq M_1 \supseteq \cdots \supseteq M_j \supseteq \cdots .$$

with the property that $\mathfrak{q}M_j \subseteq M_{j+1}$ for each $j \geq 0$. In the present work we consider only *good q-filtrations* of M : this means that $M_{j+1} = \mathfrak{q}M_j$ for all sufficiently large j. A good q-filtration is also called a *stable q-filtration*. For example, the q-adic filtration on M defined by $M_j := \mathfrak{q}^j M$ is clearly a good q-filtration.

We define the *associated graded ring* of A with respect to $\mathfrak{q}$ to be the graded ring

$$gr_{\mathfrak{q}}(A) = \bigoplus_{j\geq 0}(\mathfrak{q}^j/\mathfrak{q}^{j+1}).$$

Given a q-filtration $\mathbb{M} = \{M_j\}$ on the module M, we consider the *associated graded module* of M with respect to $\mathbb{M}$

$$gr_{\mathbb{M}}(M) := \bigoplus_{j\geq 0}(M_j/M_{j+1})$$

and for any $\bar{a} \in \mathfrak{q}^n/\mathfrak{q}^{n+1}, \bar{m} \in M_j/M_{j+1}$ we define $\bar{a}\,\bar{m} := \overline{am} \in M_{n+j}/M_{n+j+1}$. The assumption that $\mathbb{M}$ is a q-filtration ensures that this is well defined so that $gr_{\mathbb{M}}(M)$ has a natural structure as a graded module over the graded ring $gr_{\mathfrak{q}}(A)$.

Denote by $\lambda(*)$ the length of an A-module. If $\lambda(M/\mathfrak{q}M)$ is finite, then we can define the *Hilbert function* of the filtration $\mathbb{M}$, or of the filtered module M with respect to the filtration $\mathbb{M}$. It is the numerical function

$$H_{\mathbb{M}}(j) := \lambda(M_j/M_{j+1}).$$

In the classical case of the m-adic filtration on a local ring $(A, \mathfrak{m}, k)$ we write $H_A(n)$ and remark that it coincides with $\dim_k(\mathfrak{m}^n/\mathfrak{m}^{n+1})$.

Its generating function is the power series

$$P_{\mathbb{M}}(z) := \sum_{j\geq 0} H_{\mathbb{M}}(j)z^j.$$

which is called the *Hilbert series* of the filtration $\mathbb{M}$. By the Hilbert–Serre theorem we know that the series is of the form

$$P_{\mathbb{M}}(z) = \frac{h_{\mathbb{M}}(z)}{(1-z)^r}$$

where $h_{\mathbb{M}}(z) \in \mathbb{Z}[z]$, $h_{\mathbb{M}}(1) \neq 0$ and r is the Krull dimension of M. The polynomial $h_{\mathbb{M}}(z)$ is called the *h-polynomial* of $\mathbb{M}$.

This implies that, for $n \gg 0$

$$H_{\mathbb{M}}(n) = p_{\mathbb{M}}(n)$$

where the polynomial $p_{\mathbb{M}}(z)$ has rational coefficients, degree $r - 1$ and is called the *Hilbert polynomial* of $\mathbb{M}$.

We can write

$$p_{\mathbb{M}}(X) := \sum_{i=0}^{r-1} (-1)^i e_i(\mathbb{M}) \binom{X + r - i - 1}{r - i - 1}$$

where we denote for every integer $q \geq 0$

$$\binom{X + q}{q} := \frac{(X + q)(X + q - 1) \ldots (X + 1)}{q!}.$$

The coefficients $e_i(\mathbb{M})$ are integers which will be called the *Hilbert coefficients* of $\mathbb{M}$. In particular $e_0 = e_0(\mathbb{M}) = h_{\mathbb{M}}(1)$ is the *multiplicity* and it depends on M and on the ideal $\mathfrak{q}$.

When we consider the $\mathfrak{m}$-adic filtration in the local ring $(A, \mathfrak{m})$, the Hilbert function of A measures the minimal number of generators (denote $\mu(\ )$) of the powers of the maximal ideal. In the one-dimensional case the asymptotic value is the multiplicity e_0. It is a natural question to ask whether the Hilbert function of a one-dimensional Cohen–Macaulay ring is not decreasing. Clearly, this is the case if $gr_{\mathfrak{m}}(A)$ is Cohen–Macaulay, but this is not a necessary requirement.

Unfortunately, it can happen that $H_A(2) = \mu(\mathfrak{m}^2) < H_A(1) = \mu(\mathfrak{m})$. The first example was given by J. Herzog and R. Waldi in 1975. In 1980 F. Orecchia proved that, for all embedding dimension $v = \mu(\mathfrak{m}) \geq 5$, there exists a reduced one-dimensional local ring of embedding dimension v and decreasing Hilbert function. L. Roberts in 1982 built ordinary singularities with decreasing Hilbert function and embedding dimension at least 7. J. Sally conjectured, and J. Elias proved, that the Hilbert function of one-dimensional Cohen–Macaulay local rings of embedding dimension three is not decreasing (see [21] and [77]). Interesting problems are still open if we consider local domains. S. Kleiman proved that there is a finite number of admissible Hilbert functions for graded domains with fixed multiplicity and dimension. The analogous of Kleiman's result does not hold in the local case.

Nevertheless, V. Srinivas and V. Trivedi in [98] proved that the number of Hilbert functions of Cohen–Macaulay local rings with given multiplicity and dimension is finite (a different proof was given by M.E. Rossi, N.V. Trung and G. Valla in [79]). This is a very interesting result and it produces upper bounds on the Hilbert coefficients. If $(A, \mathfrak{m})$ is a Cohen–Macaulay local ring of dimension r and multiplicity e_0, then

$$e_i \leq e_0^{3i! - i} - 1 \text{ for all } i \geq 1.$$

(see [98, Theorem 1], [79, Corollary 4.2]). These bounds are far from being sharp, but they have some interest because very little is known about e_i with $i > 2$.

It is clear that different Hilbert functions can have the same Hilbert polynomial; but in many cases it happens that "extremal" behavior of some of the e_i forces the filtration to have a specified Hilbert function. The trivial case is when the multiplicity is one: if this happens, then A is a regular local ring and $P_A(z) = \frac{1}{(1-z)^r}$. Also the case of multiplicity 2 is easy, while the first non trivial result along this line was proved by J. Sally in [87]. If $(A, \mathfrak{m})$ is a Cohen–Macaulay local ring of dimension r and embedding dimension $v := H_A(1) = \mu(\mathfrak{m})$, we let $h := v - r$, the embedding codimension of A. It is a result of Abhyankar that a lower bound for the multiplicity e_0 is given by

$$e_0 \geq h + 1.$$

This result extends to the local Cohen–Macaulay rings the well known lower bound for the degree of a variety X in $\mathbb{P}^n$:

$$\deg X \geq \operatorname{codim} X + 1.$$

The varieties for which the bound is attained are called varieties of minimal degree and they are completely classified. In particular, they are always arithmetically Cohen–Macaulay. In the local case, Sally proved that if the equality $e_0 = h + 1$ holds, then $gr_\mathfrak{m}(A)$ is Cohen–Macaulay and $P_A(z) = \frac{1+hz}{(1-z)^r}$.

The next case, varieties satisfying deg $X = \operatorname{codim} X + 2$, is considerably more difficult. In particular such varieties are not necessarily arithmetically Cohen–Macaulay. Analogously, in the case $e_0 = h + 2$, in [91] it was shown that $gr_\mathfrak{m}(A)$ is not necessarily Cohen–Macaulay, the exceptions lie among the local rings of maximal Cohen–Macaulay type $\tau(A) = e_0 - 2$. In the same paper Sally made the conjecture that, in the critical case, the depth of $gr_\mathfrak{m}(A)$ is at least $r - 1$. This conjecture was proved in [115] and [80] by using deep properties of the Ratliff–Rush filtration on the maximal ideal of A. Further in [80], all the possible Hilbert functions have been described: they are of the form

$$P_A(z) = \frac{1 + hz + z^s}{(1-z)^r}$$

where $2 \leq s \leq h + 1$.

The next case, when $e_0 = h + 3$, is more complicated and indeed still largely open. J. Sally, in another paper, see [89], proved that if A is Gorenstein and $e_0 = h + 3$, then $gr_\mathfrak{m}(A)$ is Cohen–Macaulay and

$$P_A(z) = \frac{1 + hz + z^2 + z^3}{(1-z)^r}.$$

If the Cohen–Macaulay type $\tau(A)$ is bigger than 1, then $gr_\mathfrak{m}(A)$ is no longer Cohen–Macaulay. Nevertheless, if $\tau(A) < h$, in [82] the authors proved that depth$(gr_\mathfrak{m}(A))$ $\geq r - 1$ and the Hilbert series is given by

$$P_A(z) = \frac{1 + hz + z^2 + z^s}{(1-z)^r}$$

where $2 \leq s \leq \tau(A) + 2$. This gives a new and shorter proof of the result of Sally, and it points to the remaining open question: what are the possible Hilbert functions for a Cohen–Macaulay r-dimensional local ring with $e_0 = h + 3$ and $\tau(A) \geq h$?

It is clear that, moving away from the minimal value of the multiplicity, things soon become very difficult, and we do not have any idea what are the possible Hilbert functions of a one-dimensional Cohen–Macaulay local ring. The conjecture made by M.E. Rossi, that this function is non decreasing when A is Gorenstein, is very much open, even for coordinate rings of monomial curves.

Intuitively, involving the higher Hilbert coefficients should give stronger results. Indeed if $(A, \mathfrak{m})$ is Cohen–Macaulay and we consider the $\mathfrak{m}$-adic filtration on A, then D. Northcott proved in [64] that $e_1 \geq e_0 - 1$ and, if $e_1 = e_0 - 1$, then $P_A(z) = \frac{1+hz}{(1-z)^r}$ while if $e_1 = e_0$ then $P_A(z) = \frac{1+hz+z^2}{(1-z)^r}$.

Results of this kind are quite remarkable because, in principle, e_0 and e_1 give only partial information on the Hilbert polynomial and asymptotic information on the Hilbert function. The first coefficient, e_0, is the multiplicity and, due to its geometric meaning, has been studied very deeply. This integer, e_1, has been recently interpreted in [68] as a tracking number of the Rees algebra of A in the set of all such algebras with the same multiplicity. Under various circumstances, it is also called the *Chern number* of the local ring A. An interesting list of questions and conjectural statements about the values of e_1 for filtrations associated to an $\mathfrak{m}$-primary ideal of a local ring A had been presented in a recent paper by W. Vasconcelos (see [111]). A surprising result by L. Ghezzi, S. Goto, J. Hong, K. Ozeki, T.T. Phuong, W. Vasconcelos characterizes the Cohen–Macaulayness of an unmixed local ring A in terms of the vanishing of the e_1 of a system of parameters (see [28]). They proved that an unmixed local ring A is Cohen–Macaulay if and only if $e_1(J) = 0$ for some parameter ideal J (with respect to the J-adic filtration). The analogous investigation for the Buchsbaumness was discussed by S. Goto and K. Ozeki in [35]. Under suitable conditions they proved that a local ring A of dimension ≥ 2 is Buchsbaum if and only if the first Hilbert coefficients $e_1(J)$ are constant and independent of the choice of parameter ideals J in A.

Over the past few years several papers have appeared which extend classical results on the theory of Hilbert functions of Cohen–Macaulay local rings to the case of a filtration of a module. Very often, because of this increased generality, deep obstructions arise which can be overcome only by bringing new ideas to bear. Instead, in this paper we illustrate how a suitable and natural recasting of the main basic tools of the classical theory is often enough to obtain the required extensions.

More precisely what one needs is to make available in the generalized setting a few basic tools of the classical theory, such as *superficial sequences, the Valabrega–Valla criterion, Sally's machine, Singh's formula*.

Once these fundamental results have been established, the approach followed in this monograph gives a simple and clean method which applies uniformly to many cases.

In this way we make use of the usual machinery to get easier proofs, extensions of known results as well as numerous entirely new results. We mention two nice examples of this philosophy:

1. The problem of the existence of elements which are superficial simultaneously with respect to a finite number of q-filtrations on the same module has a natural solution in the module-theoretical approach (see Remark 1.1), while it is rather complicated in the ring-theoretical setting (see [31, Lemma 2.3]).

2. In the literature two different definitions of minimal multiplicity are given (see Sect. 2.1). Here they are unified, being just instances of the more general concept of minimal multiplicity with respect to different filtrations of the same ring.

The notion of *superficial element* is a fundamental tool in our work. The original definition was given by Zariski and Samuel [117, p. 285]. There it is shown how to use this concept for devising proofs by induction and reducing problems to lower dimensional ones. We are concerned only with the purely algebraic meaning of this notion, even if superficial elements play an important role also in Singularity Theory, as shown by R. Bondil and Le Dung Trang in [4] and [5].

We know that superficial sequences of order 1 always exist if the residue field is infinite, a condition which is not so restrictive. We make a lot of use of this, often reducing a problem to the one-dimensional case where things are much easier.

The main consequence of this strategy is that our arguments are quite elementary and, for example, we are able to avoid the more sophisticated homological methods used in other papers.

The extension of the theory to the case of general filtrations on a module has one more important motivation. Namely, we have interesting applications to the study of graded algebras which are not associated to a filtration. Here we have in mind the Symmetric algebra $S_A(\mathfrak{q})$, the Fiber cone $F_{\mathfrak{m}}(\mathfrak{q})$ and the Sally-module $S_J(\mathfrak{q})$ of an ideal $\mathfrak{q}$ of A with respect to a minimal reduction J. These graded algebras have been studied for their intrinsic interest; however, since the rich theory of filtrations apparently does not apply, new and complicated methods have been developed.

We show here that each of these algebras fits into certain short exact sequences together with algebras associated to filtrations. Hence we can study the Hilbert function and the depth of these algebras with the aid of the know-how we got in the case of a filtration.

This strategy has been already used in [46] to study the depth of the Symmetric Algebra $S_A(\mathfrak{m})$ of the maximal ideal $\mathfrak{m}$ of a local ring A. Also, in [17, 18], T. Cortadellas and S. Zarzuela used similar ideas to study the Cohen–Macaulayness of the Fiber cone.

In the last two chapters, we present selected results from the recent literature on the Fiber cone $F_{\mathfrak{m}}(\mathfrak{q})$ and the Sally-module $S_J(\mathfrak{q})$. We have chosen not to pursue here the study of the Symmetric algebra of the maximal ideal of a local ring, even if we think it could be interesting and fruitful to apply the ideas of this paper also to that problem.

In developing this work, one needs to consider filtrations on modules which are not necessarily Cohen–Macaulay. This opens up a new and interesting terrain because most of the research done on Hilbert functions has been carried out in the framework of Cohen–Macaulay local rings. Recently, S. Goto, K. Nishida, A. Corso, W. Vasconcelos and others have discovered interesting results on the Hilbert function of general local rings. Following their methods, we tried to develop our theory as for as possible but without the strong assumption of Cohen–Macaulayness. But soon things became so difficult that we had to return quickly to the classical assumption.

We finish this introduction by giving a brief summary of each chapter. A longer description can be found at the beginning of each chapter.

In the first chapter we introduce and discuss the notion of a good q-filtration of a module over a local ring. The corresponding associated graded module is defined, and a criterion for detecting regular sequences on this module is presented. Next we define the notions of Hilbert function and polynomial of a filtration and describe the relationship with superficial elements. Finally, we give a natural upper bound for the Hilbert function of a filtration in terms of a maximal superficial sequence.

In the second chapter we give several upper bounds for the first two Hilbert coefficients of the Hilbert polynomial of a filtration; it turns out that modules which are extremal with respect to these bounds have good associated graded modules and fully determined Hilbert functions. In particular, we present a notable generalization of Northcott's classical bound. We present several results for modules which do not require Cohen–Macaulayness. Here the theory of the one-dimensional case plays a crucial role.

The third chapter deals with the third Hilbert coefficient; some upper and lower bounds are discussed. We introduce the Ratliff–Rush filtration associated to a q-adic filtration and show some applications to the study of border cases. The proofs become more sophisticated because the complex structure of local rings of dimension at least two comes into play.

In the fourth chapter we give a proof of Sally's conjecture in a very general context, thus greatly extending the classical case. Several applications to the first Hilbert coefficients are discussed. The main result of this chapter is a bound on the reduction number of a filtration which has unexpected applications.

In Chaps. 5 and 6 we explore the depth and the Hilbert coefficients of the Fiber cone and the Sally module respectively. In spite of the fact that the Fiber cone and the Sally module are not graded modules associated to a filtration, the aim of this section is to show how one can deduce their properties as a consequence of the theory on filtrations. In particular we will get short proofs of several recent results as an easy consequence of certain classical results on the associated graded rings to special filtrations.

Chapter 1
Preliminaries

In this chapter we present the basic tools of the classical theory of filtered modules, in particular we introduce the machinery we shall use throughout this work: $\mathbb{M}$-superficial elements and their interplay with Hilbert Functions, the Valabrega–Valla criterion, which is a basic tool for studying the depth of $\mathrm{gr}_{\mathbb{M}}(M)$, $\mathbb{M}$-superficial sequences for an ideal $\mathfrak{q}$ and their relevance to Sally's machine, which is a very important device for reducing dimension in questions relating to depth properties of blowing-up rings and local rings.

1.1 Notation

Let A be a commutative noetherian local ring with maximal ideal $\mathfrak{m}$ and let M be a finitely generated A-module. We will denote by $\lambda(\cdot)$ the length of an A-module. An (infinite) chain

$$M = M_0 \supseteq M_1 \supseteq \cdots \supseteq M_j \supseteq \cdots$$

where the M_n are submodules of M is called a *filtration* of M, and denoted by $\mathbb{M} = \{M_n\}$. Given an ideal $\mathfrak{q}$ in A, $\mathbb{M}$ is a $\mathfrak{q}$-*filtration* if $\mathfrak{q}M_j \subseteq M_{j+1}$ for all j, and a *good* $\mathfrak{q}$-*filtration* if $M_{j+1} = \mathfrak{q}M_j$ for all sufficiently large j. Thus for example $\{\mathfrak{q}^n M\}$ is a good $\mathfrak{q}$-filtration. In the literature a good $\mathfrak{q}$-filtration is sometimes called a stable $\mathfrak{q}$-filtration. We say that $\mathbb{M}$ is nilpotent if $M_n = 0$ for $n \gg 0$. Thus a good $\mathfrak{q}$-filtration $\mathbb{M}$ is nilpotent if and only if $\mathfrak{q} \subseteq \sqrt{Ann\, M}$.

From now on $\mathbb{M}$ will denote always a good $\mathfrak{q}$-filtration on the finitely generated A-module M.

We will assume that the ideal $\mathfrak{q}$ is proper. As a consequence $\cap_{i=0}^{\infty} M_i = \{0_M\}$. Define

$$gr_{\mathfrak{q}}(A) = \bigoplus_{j \geq 0} (\mathfrak{q}^j / \mathfrak{q}^{j+1}).$$

This is a graded ring in which the multiplication is defined as follows: if $a \in \mathfrak{q}^i$, $b \in \mathfrak{q}^j$ define $\overline{a}\overline{b}$ to be $\overline{ab}$, i.e. the image of ab in $\mathfrak{q}^{i+j}/\mathfrak{q}^{i+j+1}$. This ring is the *associated graded ring* of the ideal $\mathfrak{q}$.

M.E. Rossi and G. Valla, *Hilbert Functions of Filtered Modules*, Lecture Notes of the Unione Matematica Italiana 9, DOI 10.1007/978-3-642-14240-6_1, © Springer-Verlag Berlin Heidelberg 2010

Similarly, if M is an A-module and $\mathbb{M} = \{M_j\}$ is a q-filtration of M, define

$$gr_{\mathbb{M}}(M) = \bigoplus_{j \geq 0} (M_j/M_{j+1})$$

which is a graded $gr_{\mathfrak{q}}(A)$-module in a natural way. It is called the *associated graded module* to the q-filtration $\mathbb{M} = \{M_n\}$.

To avoid triviality we shall assume that $gr_{\mathbb{M}}(M)$ is not zero or equivalently $M \neq 0$. Each element $a \in A$ has a natural image, denoted by $a^* \in gr_{\mathfrak{q}}(A)$ and called initial form of a with respect to $\mathbb{M}$. If $a = 0$, then $a^* = 0$, otherwise $a^* = \overline{a} \in \mathfrak{q}^t/\mathfrak{q}^{t+1}$ where t is the unique integer such that $a \in \mathfrak{q}^t$, $a \notin \mathfrak{q}^{t+1}$.

If N is a submodule of M, by the Artin-Rees Lemma, the collection $\{N \cap M_j \mid j \geq 0\}$ is a good q-filtration of N. Since

$$(N \cap M_j)/(N \cap M_{j+1}) \simeq (N \cap M_j + M_{j+1})/M_{j+1}$$

$gr_{\mathbb{M}}(N)$ is a graded submodule of $gr_{\mathbb{M}}(M)$.

On the other hand it is clear that $\{(N + M_j)/N \mid j \geq 0\}$ is a good q-filtration of M/N which we denote by $\mathbb{M}/N$ and we call quotient filtration. These graded modules are related by the graded isomorphism

$$gr_{\mathbb{M}/N}(M/N) \simeq gr_{\mathbb{M}}(M)/gr_{\mathbb{M}}(N).$$

If $a_1, \ldots, a_r$ are elements in $\mathfrak{q} \setminus \mathfrak{q}^2$ and $I = (a_1, \ldots, a_r)$, it is clear that

$$[(a_1^*, \ldots, a_r^*)gr_{\mathbb{M}}(M)]_j = (IM_{j-1} + M_{j+1})/M_{j+1}$$

for each $j \geq 1$. By the Artin-Rees Lemma one immediately gets that the following conditions are equivalent:

(1) $gr_{\mathbb{M}/IM}(M/IM) \simeq gr_{\mathbb{M}}(M)/(a_1^*, \ldots, a_r^*)gr_{\mathbb{M}}(M)$.

(2) $gr_{\mathbb{M}}(IM) = (a_1^*, \ldots, a_r^*)gr_{\mathbb{M}}(M)$.

(3) $IM \cap M_j = IM_{j-1} \ \forall j \geq 1$.

An interesting case in which the above equalities hold is when the elements $a_1^*, \ldots, a_r^*$ form a regular sequence on $gr_{\mathbb{M}}(M)$. For example, if $r = 1$ and $I = (a)$, then $\overline{a} \in \mathfrak{q}/\mathfrak{q}^2$ is regular on $gr_{\mathbb{M}}(M)$ if and only if the map

$$M_{j-1}/M_j \xrightarrow{\overline{a}} M_j/M_{j+1}$$

is injective for every $j \geq 1$. This is equivalent to the equalities $M_{j-1} \cap (M_{j+1} : a) = M_j$ for every $j \geq 1$. An easy computation shows the following result.

Lemma 1.1. *Let $a \in \mathfrak{q}$. The following conditions are equivalent:*

(1) $\overline{a} \in \mathfrak{q}/\mathfrak{q}^2$ is a regular element on $gr_{\mathbb{M}}(M)$.

(2) $M_{j-1} \cap (M_{j+1} : a) = M_j$ *for every $j \geq 1$.*

(3) a is a regular element on M and $aM \cap M_j = aM_{j-1}$ for every $j \geq 1$.

(4) $M_{j+1} : a = M_j$ *for every $j \geq 0$.*

This result leads us to the Valabrega–Valla criterion, a tool which has been very useful in the study of the depth of blowing-up rings (see [105]).

Many authors have discussed this topic recently. For example in [53] Huckaba and Marley gave an extension of the classical result to the case of filtrations of ideals, by giving a completely new proof based on some deep investigation of a modified Koszul complex.

Instead, in [69], Puthenpurakal extended the result to the case of q-adic filtrations of a module by using the device of idealization of a module and then applying the classical result.

This would suggest that the original proof does not work in the more general setting. But, after looking at it carefully, we can say that, in order to prove the following very general statement, one does not need any new idea: a straightforward adaptation of the dear old proof does the job.

Theorem 1.1 (Valabrega–Valla). *Let $a_1, \ldots, a_r$ be elements in $\mathfrak{q} \setminus \mathfrak{q}^2$ and I the ideal they generate. Then $a_1^*, \ldots, a_r^*$ form a regular sequence on $gr_\mathbb{M}(M)$ if and only if $a_1, \ldots, a_r$ form a regular sequence on M and $IM \cap M_j = IM_{j-1} \ \forall j \geq 1$.*

1.2 Superficial Elements

A fundamental tool in local algebra is the notion of superficial element. This notion goes back to P. Samuel and our methods are also related to the construction given by Zariski and Samuel [117, p. 296].

Definition 1.1. An element $a \in \mathfrak{q}$, is called $\mathbb{M}$-*superficial* for $\mathfrak{q}$ if there exists a non-negative integer c such that

$$(M_{n+1} :_M a) \cap M_c = M_n$$

for every $n \geq c$.

For every $a \in \mathfrak{q}$ and $n \geq c$, M_n is contained in $(M_{n+1} :_M a) \cap M_c$. Then it is the other inclusion that makes superficial elements special. It is clear that if $\mathbb{M}$ is nilpotent, then every element is superficial. If the length $\lambda(M/\mathfrak{q}M)$ is finite, then $\mathbb{M}$ is nilpotent if and only if $\dim M = 0$. Hence in the following, when we deal with superficial elements, *we shall assume that* $\dim M \geq 1$.

If this is the case, as a consequence of the definition, we deduce that $\mathbb{M}$-superficial elements a for $\mathfrak{q}$ have order one, that is $a \in \mathfrak{q} \setminus \mathfrak{q}^2$. With a slight modification of the given definition, superficial elements can be introduced of every order, but in the following we need superficial elements of order 1 because they have a better behaviour for studying Hilbert functions.

Hence superficial element always means a superficial element of order 1. It is well known that superficial elements do not always exist, but their existence is guaranteed if the residue field is infinite (see [56, Proposition 8.5.7]). By passing, if needed, to the faithfully flat extension $A[x]_{\mathfrak{m}A[x]}$ (x is a variable over A) we may assume that the residue field is infinite.

If $\mathfrak{q}$ contains a regular element on M, it is easy to see that every $\mathbb{M}$-superficial element of $\mathfrak{q}$ is regular on M.

Given A-modules M and N, let $\mathbb{M}$ and $\mathbb{N}$ be good $\mathfrak{q}$-filtrations of M and N respectively. We define a new filtration as follows

$$\mathbb{M} \oplus \mathbb{N}: \quad M \oplus N \supseteq M_1 \oplus N_1 \supseteq \cdots \supseteq M_n \oplus N_n \supseteq \ldots$$

It is easy to see that $\mathbb{M} \oplus \mathbb{N}$ is a good $\mathfrak{q}$-filtration on the A-module $M \oplus N$. Of course this construction can be extended to any finite number of modules.

The following remark is due to David Conti in his thesis (see [11]).

Remark 1.1. Let $\mathbb{M}_1, \ldots, \mathbb{M}_n$ be $\mathfrak{q}$-filtrations of M and let $a \in \mathfrak{q}$. Then a is $\mathbb{M}_1 \oplus \cdots \oplus \mathbb{M}_n$-superficial for $\mathfrak{q}$ if and only if a is $\mathbb{M}_i$-superficial for $\mathfrak{q}$ for every $i = 1, \ldots, n$.

This result is an easy consequence of the good behaviour of intersection and colon of ideals with respect to direct sum of modules. As a consequence we deduce that, if the residue field is infinite, we can always find an element $a \in \mathfrak{q}$ which is superficial for a finite number of $\mathfrak{q}$-filtrations on M. As mentioned in the introduction, we want to apply the general theory of the filtrations on a module to the study of certain blowing-up rings which are not necessarily associated graded rings to a single filtration. Since they are related to different filtrations, the above remark will be relevant in our approach.

David Conti also remarked that $\mathbb{M}$-superficial elements of order $s \geq 1$ for $\mathfrak{q}$ can be seen as superficial elements of order 1 for a suitable filtration which is strictly related to $\mathbb{M}$. Let $\mathbb{N}$ be the $\mathfrak{q}^s$-filtration:

$$M \oplus M_1 \oplus \cdots \oplus M_{s-1} \supseteq M_s \oplus M_{s+1} \oplus \cdots \oplus M_{2s-1} \supseteq M_{2s} \oplus M_{2s+1} \oplus \cdots \oplus M_{3s-1} \supseteq \ldots$$

Then it is easy to see that an element a is $\mathbb{M}$-superficial of degree s for $\mathfrak{q}$ if and only if a is $\mathbb{N}$-superficial of degree one for $\mathfrak{q}^s$. This remark could be useful in studying properties which behave well with the direct sum.

We give now equivalent conditions for an element to be $\mathbb{M}$-superficial for $\mathfrak{q}$. Our development of the theory of superficial elements is basically the same as that given by Kirby in [60] for the case $M = A$ and $\{M_j\} = \{\mathfrak{q}^j\}$ the $\mathfrak{q}$-adic filtration on A. If there is no confusion, we let

$$G := gr_{\mathbb{M}}(M), \quad Q := \bigoplus_{j \geq 1} (\mathfrak{q}^j / \mathfrak{q}^{j+1}).$$

Let

$$\{0_M\} = P_1 \cap \cdots \cap P_r \cap P_{r+1} \cap \cdots \cap P_s$$

be an irredundant primary decomposition of $\{0_M\}$ and let $\{\wp_i\} = Ass(M/P_i)$. Then $Ass(M) = \{\wp_1, \ldots, \wp_s\}$ and $\wp_i = \sqrt{0 : (M/P_i)}$.

We may assume $\mathfrak{q} \nsubseteq \wp_i$ for $i = 1, \ldots, r$ and $\mathfrak{q} \subseteq \wp_i$ for $i = r+1, \ldots, s$. Then we let

$$N := P_1 \cap \cdots \cap P_r.$$

It is clear that

$$N = \{x \in M \mid \exists\, n, \mathfrak{q}^n x = 0_M\}$$

and that $N \cap M_j = \{0\}$ for all large j and $Ass(M/N) = \bigcup_{i=1}^r \wp_i$.

Similarly we denote by H the homogeneous submodule of G consisting of the elements $\alpha \in G$ such that $Q^n \alpha = 0_G$, hence

$$H = \{\alpha \in G \mid \exists\, n, Q^n \alpha = 0_G\}.$$

If

$$\{0_G\} = T_1 \cap \cdots \cap T_m \cap T_{m+1} \cap \cdots \cap T_l$$

is an irredundant primary decomposition of $\{0_G\}$ and we let $\{\mathfrak{P}_i\} = Ass(G/T_i)$, then $Ass(G) = \{\mathfrak{P}_1, \ldots, \mathfrak{P}_l\}$ and $\mathfrak{P}_i = \sqrt{0 : (G/T_i)}$.

Further, if we assume that $Q \nsubseteq \mathfrak{P}_i$ for $i = 1, \ldots, m$ and $Q \subseteq \mathfrak{P}_i$ for $i = m+1, \ldots, l$, then

$$H = T_1 \cap \cdots \cap T_m$$

and $Ass(G/H) = \bigcup_{i=1}^m \mathfrak{P}_i$.

Theorem 1.2. *Let $a \in \mathfrak{q} \backslash \mathfrak{q}^2$, the following conditions are equivalent:*

(1) a is $\mathbb{M}$-superficial for $\mathfrak{q}$.

(2) $a^* \notin \bigcup_{i=1}^m \mathfrak{P}_i$.

(3) $H :_G a^* = H$.

(4) $N : a = N$ and $M_{j+1} \cap aM = aM_j$ for all large j.

(5) $(0 :_G a^*)_j = 0$ for all large j.

(6) $M_{j+1} : a = M_j + (0 :_M a)$ and $M_j \cap (0 :_M a) = 0$ for all large j.

We note with [117] that if the residue field $A/\mathfrak{m}$ is infinite, condition (2) of the above theorem ensures the existence of $\mathbb{M}$-superficial elements for $\mathfrak{q}$ as we have already mentioned. Moreover condition (5) says that a is $\mathbb{M}$-superficial for $\mathfrak{q}$ if and only if a^* is homogeneous *filter-regular* in G. We may refer to [101, 102, 104], concerning the definition and the properties of the homogeneous filter-regular elements. Filter-regular elements were also introduced in the local contest by N.T. Cuong, P. Schenzel and N.V. Trung in [19]. One of the main results in [19] says that a local ring is generalized Cohen–Macaulay if and only if every system of parameters is filter-regular. The notion of a filter-regular element in a local ring is weaker than superficial element and in general it does not behave well with Hilbert functions.

A superficial element is filter-regular, but the converse does not hold. It is enough to recall that if a is M-regular, then accordingly with [19], a is a filter-regular element, but it is not necessarily a superficial element.

A sequence of elements $a_1, \ldots, a_r$ will be called a $\mathbb{M}$-*superficial sequence for* q if, for $i = 1, \ldots, r$, a_i is an $(\mathbb{M}/(a_1, \ldots, a_{i-1})M)$-superficial element for q.

In order to prove properties of superficial sequences often we can argue by induction on the number of elements, the above theorem giving the first step of the induction. For example, since $\operatorname{depth}_q(M) \geq 1$ implies $N = 0$, condition (4) gives the following result. Here $\operatorname{depth}_q(M)$ denotes the common cardinality of all the maximal M-regular sequences of elements in q.

The following lemmas will be crucial devices through the whole paper.

Lemma 1.2. *Let $a_1, \ldots, a_r$ be an $\mathbb{M}$-superficial sequence for q. Then $a_1, \ldots, a_r$ is a regular sequence on M if and only if $\operatorname{depth}_q(M) \geq r$.*

In the same way, since $\operatorname{depth}_Q(gr_\mathbb{M}(M)) \geq 1$ implies $H = 0$, condition 3 and Theorem 1.1 give the following result which shows the relevance of superficial elements in the study of the depth of blowing-up rings.

Lemma 1.3. *Let $a_1, \ldots, a_r$ be an $\mathbb{M}$-superficial sequence for q. Then $a_1^*, \ldots, a_r^*$ is a regular sequence on $gr_\mathbb{M}(M)$ if and only if $\operatorname{depth}_Q(gr_\mathbb{M}(M)) \geq r$.*

Now we come to *Sally's machine* or *Sally's descent*, a very important device for reducing dimension in questions relating to depth properties of blowing-up rings.

Lemma 1.4. (Sally's machine) *Let $a_1, \ldots, a_r$ be an $\mathbb{M}$-superficial sequence for q and I the ideal they generate. Then $\operatorname{depth}_Q(gr_{\mathbb{M}/IM}(M/IM)) \geq 1$, if and only if $\operatorname{depth}_Q(gr_\mathbb{M}(M)) \geq r + 1$.*

A proof of the *if* part can be obtained by a straightforward adaptation of the original proof given by Huckaba and Marley in [53, Lemma 2.2]. The converse is an easy consequence of Lemma 1.2 and Theorem 1.1.

We will also need a property of superficial elements which seems to be neglected in the literature. It is well known that if a is M-regular, then M/aM is Cohen–Macaulay if and only if M is Cohen–Macaulay. We prove, as a consequence of the following Lemma, that the same holds for a superficial element, if the dimension of the module M is at least two.

In the following we denote by $H_q^i(M)$ the i-th local cohomology module of M with respect to q. We know that $H_q^0(M) := \cup_{j \geq 0}(0 :_M q^j) = 0 :_M q^t$ for every $t \gg 0$; further $\min\{i \mid H_q^i(M) \neq 0\} = \operatorname{depth}_q(M)$.

Lemma 1.5. *Let a be an $\mathbb{M}$-superficial element for q and let $j \geq 1$. Then we have $\operatorname{depth}_q(M) \geq j + 1$ if and only if $\operatorname{depth}_q(M/aM) \geq j$.*

Proof. Let $\operatorname{depth}_q(M) \geq j + 1$; then $\operatorname{depth}_q(M) > 0$ so that a is M-regular. This implies $\operatorname{depth}_q(M/aM) = \operatorname{depth}_q(M) - 1 \geq j + 1 - 1 = j$.

Let us assume now that $\operatorname{depth}_q(M/aM) \geq j$. Since $j \geq 1$, this implies $H_q^0(M/aM) = 0$. Hence $H_q^0(aM) = H_q^0(M)$, so that $H_q^0(M) \subseteq aM$. We claim that

$H_{\mathfrak{q}}^0(M) = aH_{\mathfrak{q}}^0(M)$. If this is the case, then, by Nakayama, we get $H_{\mathfrak{q}}^0(M) = 0$ which implies $\mathrm{depth}(M) > 0$, so that a is M-regular. Hence

$$\mathrm{depth}_{\mathfrak{q}}(M) = \mathrm{depth}_{\mathfrak{q}}(M/aM) + 1 \geq j + 1,$$

as wanted.

Let us prove the claim. Suppose by contradiction that

$$aH_{\mathfrak{q}}^0(M) \subsetneq H_{\mathfrak{q}}^0(M) \subseteq aM,$$

and let $ax \in H_{\mathfrak{q}}^0(M), x \in M \setminus H_{\mathfrak{q}}^0(M)$. This means that for every $t \gg 0$ we have

$$\begin{cases} a\mathfrak{q}^t x = 0 \\ \mathfrak{q}^t x \neq 0 \end{cases}$$

We prove that this implies that a is not $\mathbb{M}$-superficial for $\mathfrak{q}$. Namely, given a positive integer c, we can find an integer $t \geq c$ and an element $d \in \mathfrak{q}^t$ such that $adx = 0$ and $dx \neq 0$. Since $\cap M_i = \{0\}$, we have $dx \in M_{j-1} \setminus M_j$ for some integer j. Now, $d \in \mathfrak{q}^t$ hence $dx \in M_t \subseteq M_c$, which implies $j \geq c$. Finally we have $dx \in M_c$, $dx \notin M_j$ and $adx = 0 \in M_{j+1}$, hence

$$(M_{j+1} :_M a) \cap M_c \supsetneq M_j.$$

The claim and the Lemma are proved. □

It is interesting to recall that the integral closure of the ideals generally behaves well reducing modulo a superficial element. For example S. Itoh [58, p. 648] proved that:

Proposition 1.1. *If $\mathfrak{q}$ is an $\mathfrak{m}$-primary ideal which is integrally closed in a Cohen–Macaulay local ring $(A, \mathfrak{m})$ of dimension $r \geq 2$, then (at least after passing to a faithfully flat extension) there exists a superficial element $a \in \mathfrak{q}$ such that $\mathfrak{q}/(a)$ is integrally closed in $A/(a)$.*

This result will be useful in proofs working by induction on r. The compatibility of integrally closed ideals with specialization by generic elements can be extended to that of modules (see [48]).

However, as we will see later, superficial elements do not behave well for Ratliff–Rush closed ideals: there exist many ideals of A all of whose powers are Ratliff–Rush closed, yet for every superficial element $a \in \mathfrak{q}$, $\mathfrak{q}/(a)$ is not Ratliff–Rush closed. Examples are given by Rossi and Swanson in [78]. Recently Puthenpurakal in [71] characterized local rings and ideals for which the Ratliff–Rush filtration *behaves well* modulo a superficial element.

It is interesting to recall that Trung and Verma in [103] introduced superficial sequences with respect to a set of ideals by working in a multigraded context.

1.3   The Hilbert Function and Hilbert Coefficients

Let $\mathbb{M}$ be a good q-filtration on the A-module M. From now on we shall require the assumption that the length of $M/\mathfrak{q}M$, which we denote by $\lambda(M/\mathfrak{q}M)$, is finite. In this case there exists an integer s such that $\mathfrak{m}^s M \subseteq (\mathfrak{q} + (0 :_A M))M$, hence (see [65]), the ideal $\mathfrak{q} + (0 :_A M)$ is primary for the maximal ideal $\mathfrak{m}$. Also the length of M/M_j is finite for all $j \geq 0$.

From now on $\mathfrak{q}$ will denote an $\mathfrak{m}$-primary ideal of the local ring $(A, \mathfrak{m})$.

In this setting we can define the *Hilbert function* of the filtration $\mathbb{M}$ or simply of the filtered module M, if there is no confusion. By definition it is the function

$$H_{\mathbb{M}}(j) := \lambda(M_j/M_{j+1}).$$

It is also useful to consider the numerical function

$$H^1_{\mathbb{M}}(j) := \lambda(M/M_{j+1}) = \sum_{i=0}^{j} H_{\mathbb{M}}(i)$$

which is called the *Hilbert–Samuel function* of the filtration $\mathbb{M}$ or of the filtered module M.

The *Hilbert series* of the filtration $\mathbb{M}$ is the power series

$$P_{\mathbb{M}}(z) := \sum_{j \geq 0} H_{\mathbb{M}}(j)z^j.$$

The power series

$$P^1_{\mathbb{M}}(z) := \sum_{j \geq 0} H^1_{\mathbb{M}}(j)z^j$$

is called the *Hilbert–Samuel series* of $\mathbb{M}$. It is clear that

$$P_{\mathbb{M}}(z) = (1 - z)P^1_{\mathbb{M}}(z).$$

By the Hilbert–Serre theorem, see for example [6], we can write

$$P_{\mathbb{M}}(z) = \frac{h_{\mathbb{M}}(z)}{(1 - z)^r}$$

where $h_{\mathbb{M}}(z) \in \mathbb{Z}[z]$, $h_{\mathbb{M}}(1) \neq 0$ and r is the dimension of M.

The polynomial $h_{\mathbb{M}}(z)$ is called the *h-polynomial* of $\mathbb{M}$ and we clearly have

$$P^1_{\mathbb{M}}(z) = \frac{h_{\mathbb{M}}(z)}{(1 - z)^{r+1}}.$$

An easy computation shows that if we let for every $i \geq 0$

$$e_i(\mathbb{M}) := \frac{h_\mathbb{M}^{(i)}(1)}{i!}$$

then for $n \gg 0$ we have

$$H_\mathbb{M}(n) = \sum_{i=0}^{r-1} (-1)^i e_i(\mathbb{M}) \binom{n+r-i-1}{r-i-1}.$$

The polynomial

$$p_\mathbb{M}(X) := \sum_{i=0}^{r-1} (-1)^i e_i(\mathbb{M}) \binom{X+r-i-1}{r-i-1}$$

has rational coefficients and is called the *Hilbert polynomial* of $\mathbb{M}$; it satisfies the equality

$$H_\mathbb{M}(n) = p_\mathbb{M}(n)$$

for $n \gg 0$. The integers $e_i(\mathbb{M})$ are called the *Hilbert coefficients* of $\mathbb{M}$ or of the filtered module M with respect to the filtration $\mathbb{M}$.

In particular $e_0(\mathbb{M})$ is the *multiplicity* of $\mathbb{M}$ and, by Proposition 11.4 in [2] (also Proposition 4.6.5 in [6]) $e_0(\mathbb{M}) = e_0(\mathbb{N})$, for every pair of good q-filtrations $\mathbb{M}$ and $\mathbb{N}$ of M.

Since $P_\mathbb{M}^1(z) = \frac{h_\mathbb{M}(z)}{(1-z)^{r+1}}$ the polynomial

$$p_\mathbb{M}^1(X) := \sum_{i=0}^{r} (-1)^i e_i(\mathbb{M}) \binom{X+r-i}{r-i}$$

satisfies the equality

$$H_\mathbb{M}^1(n) = p_\mathbb{M}^1(n)$$

for $n \gg 0$ and is called the *Hilbert–Samuel polynomial* of M.

In the following we will write the h-polynomial of M in the form

$$h_\mathbb{M}(z) = h_0(\mathbb{M}) + h_1(\mathbb{M})z + \cdots + h_s(\mathbb{M})z^s,$$

so that the integers $h_i(\mathbb{M})$ are well defined for every $i \geq 0$ and we have

$$e_i(\mathbb{M}) = \sum_{k \geq i} \binom{k}{i} h_k(\mathbb{M}). \tag{1.1}$$

Finally we remark that if M is Artinian, then $e_0(\mathbb{M}) = \lambda(M)$.

In the case of the q-adic filtration on the ring A, we will denote by $H_q(j)$ the Hilbert function, by $P_q(z)$ the Hilbert series, and by $e_i(q)$ the Hilbert coefficients of the q-adic filtration. In the case of the q-adic filtration on a module M, we will replace q with qM in the above notations.

The following result is called *Singh's formula* because the corresponding equality in the classical case was obtained by Singh in [97]. See also [117, Lemma 3, Chap. 8] for the corresponding equality with Hilbert polynomials.

Lemma 1.6. *Let* $a \in$ q; *then for every* $j \geq 0$ *we have*

$$H_{\mathbb{M}}(j) = H^1_{\mathbb{M}/aM}(j) - \lambda(M_{j+1} : a/M_j).$$

Proof. The proof is an easy consequence of the following exact sequence:

$$0 \to (M_{j+1} : a)/M_j \to M/M_j \xrightarrow{a} M/M_{j+1} \to M/(M_{j+1} + aM) \to 0.$$

$\square$

We remark that, by Singh's formula, for every $a \in$ q we have

$$P_{\mathbb{M}}(z) \leq P^1_{\mathbb{M}/aM}(z) \tag{1.2}$$

It is thus interesting to consider elements $a \in$ q such that $P_{\mathbb{M}}(z) = P^1_{\mathbb{M}/aM}(z)$. By the above formula and Lemma 1.1, these are exactly the elements $a \in$ q such that $\bar{a} \in$ q/q^2 is regular over $gr_{\mathbb{M}}(M)$.

As a corollary of Singh's formula we get a number of useful properties of superficial elements.

Proposition 1.2. *Let* a *be an* $\mathbb{M}$-*superficial element for* q *and let* $r = \dim M \geq 1$. *Then we have:*

(1) $\dim(M/aM) = r - 1$.

(2) $e_j(\mathbb{M}) = e_j(M/aM)$ *for every* $j = 0, \ldots, r - 2$.

(3) $e_{r-1}(\mathbb{M}/aM) = e_{r-1}(\mathbb{M}) + (-1)^{r-1}\lambda(0 : a)$.

(4) *There exists an integer* j *such that for every* $n \geq j - 1$ *we have* $e_r(\mathbb{M}/aM) = e_r(\mathbb{M}) + (-1)^r \left(\sum_{i=0}^n \lambda(M_{i+1} : a/M_i) - (n+1)\lambda(0 : a) \right)$.

(5) a^* *is a regular element on* $gr_{\mathbb{M}}(M)$ *if and only if* $P_{\mathbb{M}}(z) = P^1_{\mathbb{M}/aM}(z) = \frac{P_{M/aM}(z)}{1-z}$ *if and only if* a *is* M-*regular and* $e_r(\mathbb{M}) = e_r(\mathbb{M}/aM)$.

Proof. By Lemma 1.6 we have

$$P_{\mathbb{M}}(z) = P^1_{\mathbb{M}/aM}(z) - \sum_{i \geq 0} \lambda(M_{i+1} : a/M_i)z^i.$$

Since a is superficial, by Theorem 1.2, part 6, there exists an integer j such that for every $n \geq j$ we have

$$M_{n+1} : a/M_n = M_n + (0 : a)/M_n \simeq (0 : a)/(0 : a) \cap M_n = (0 : a).$$

Hence we can write

$$P_{\mathbb{M}}(z) = P^1_{\mathbb{M}/a\mathbb{M}}(z) - \sum_{i=0}^{j-1} \lambda(M_{i+1} : a/M_i)z^i - \frac{\lambda(0 : a)z^j}{(1-z)}.$$

This proves (1) (actually (1) follows also from Theorem 1.2(4)) and, as a consequence, we get

$$h_{\mathbb{M}}(z) = h_{\mathbb{M}/a\mathbb{M}}(z) - (1-z)^r \left(\sum_{i=0}^{j-1} \lambda(M_{i+1} : a/M_i)z^i \right) - (1-z)^{r-1}\lambda(0 : a)z^j.$$

This gives easily (2), (3) and (4), while (5) follows from (4) and Lemma 1.1.   □

As we can see in the proof of the above result, if a is an $\mathbb{M}$-superficial element, there exists an integer j such that $M_{n+1} : a/M_n = (0 : a)$ for every $n \geq j$. Hence, Proposition 1.2 part 4 can be rewritten as follows

$$e_r(\mathbb{M}/a\mathbb{M}) = e_r(\mathbb{M}) + (-1)^r \left(\sum_{i=0}^{j-1} \lambda(M_{i+1} : a/M_i) - j\lambda(0 : a) \right).$$

For example let $A = k[[X,Y,Z]]/(X^3, X^2Y^3, X^2Z^4) = k[[x,y,z]]$. We have $r = \dim(A) = 2$ and if we consider on $M = A$ the $\mathfrak{m}$-adic filtration $\mathbb{M} = \{\mathfrak{m}^j\}$, then y is an $\mathbb{M}$-superficial element for $\mathfrak{q} = \mathfrak{m}$. We have

$$P_{\mathbb{M}}(z) = \frac{1 + z + z^2 - z^5 - z^6 + z^9}{(1-z)^2}$$

so that

$$e_0(\mathbb{M}) = 2, \quad e_1(\mathbb{M}) = 1, \quad e_2(\mathbb{M}) = 12.$$

Also it is clear that

$$\lambda(\mathfrak{m}^{n+1} : y/\mathfrak{m}^n) = \begin{cases} 0 & \text{for } n = 0, \ldots, 4, \\ 1 & \text{for } n = 5, \\ 2 & \text{for } n = 6, \\ 3 & \text{for } n = 7, \\ 4 = \lambda(0 : y) & \text{for } n \geq 8 \end{cases}$$

so that $j = 8$ in the above proposition. Hence, by using Proposition 1.2,

$$e_0(\mathbb{M}/(y)) = 2, \quad e_1(\mathbb{M}/(y)) = -3, \quad e_2(\mathbb{M}/(y)) = -14$$

Notice that

$$P_{\mathbb{M}/(y)}(z) = \frac{1 + z + z^2 - z^6}{1 - z}.$$

Remark 1.2. By Proposition 1.2, Theorem 1.2 and Singh's formula, if $a \in \mathfrak{q}$ is an element which is M-regular, then

$$a \text{ is } \mathbb{M}\text{-superficial for } \mathfrak{q} \iff e_i(\mathbb{M}) = e_i(\mathbb{M}/aM) \text{ for every } i = 0,\ldots,r-1.$$

Thus we get a useful criterion for checking whether an element is superficial or not. From the computational point of view it reduces infinitely many conditions, coming a priori from the definition of superficial element, to the computation of the Hilbert polynomials of $\mathbb{M}$ and $\mathbb{M}/aM$ which is often available, for example, in the CoCoA system.

An $\mathbb{M}$-superficial sequence $a_1,\ldots,a_r$ for $\mathfrak{q}$ is called a *maximal $\mathbb{M}$-superficial sequence* if $\mathbb{M}/(a_1,\ldots,a_r)M$ is nilpotent, but $\mathbb{M}/(a_1,\ldots,a_{r-1})M$ is not nilpotent. By Proposition 1.2, a superficial sequence is in particular a system of parameters, hence if $\dim M = r$, every $\mathbb{M}$-superficial sequence $a_1,\ldots,a_r$ for $\mathfrak{q}$ is a maximal superficial sequence.

In our approach maximal $\mathbb{M}$-superficial sequences will play a fundamental role. We also remark that they are strictly related to minimal reductions.

Let $\mathbb{M}$ be a good $\mathfrak{q}$-filtration of M and $J \subseteq \mathfrak{q}$ an ideal. Then J is an $\mathbb{M}$-reduction of $\mathfrak{q}$ if $M_{n+1} = JM_n$ for $n \gg 0$. We say that J is a *minimal $\mathbb{M}$-reduction* of $\mathfrak{q}$ if it is minimal with respect to the inclusion.

If $\mathfrak{q}$ is $\mathfrak{m}$-primary and the residue field is infinite, there is a complete correspondence between maximal $\mathbb{M}$-superficial sequences for $\mathfrak{q}$ and minimal $\mathbb{M}$-reductions of $\mathfrak{q}$. Every minimal $\mathbb{M}$-reduction J of $\mathfrak{q}$ can be generated by a maximal $\mathbb{M}$-superficial sequence, conversely the ideal generated by a maximal $\mathbb{M}$-superficial sequence is a minimal $\mathbb{M}$-reduction of $\mathfrak{q}$ (see [56] and [11]).

Notice that if we consider for example $\mathfrak{q} = (x^7, x^6 y, x^3 y^4, x^2 y^5, y^7)$ in the power series ring $k[[x,\ y]]$, then $J = (x^7, y^7)$ is a minimal reduction of $\mathfrak{q}$, but $\{x^7, y^7\}$ is not a superficial sequence, in particular x^7 is not superficial for $\mathfrak{q}$. Nevertheless $\{x^7 + y^7,\ x^7 - y^7\}$ is a minimal system of generators of J and it is a superficial sequence for $\mathfrak{q}$.

In this presentation we prefer to handle $\mathbb{M}$-superficial sequences with respect to minimal reductions because they have a better behaviour for studying Hilbert functions and Hilbert coefficients via Proposition 1.2.

1.4 Maximal Hilbert Functions

Superficial sequences play an important role in the following result where maximal Hilbert functions are described. The result was proved in the classical case in [83, Theorem 2.2] and here is extended to the filtrations of a module which is not necessarily Cohen–Macaulay.

Theorem 1.3. *Let M be a module of dimension $r \geq 1$ and let J be the ideal generated by a maximal $\mathbb{M}$-superficial sequence for $\mathfrak{q}$. Then*

$$P_{\mathbb{M}}(z) \leq \frac{\lambda(M/M_1) + \lambda(M_1/JM)z}{(1-z)^r}.$$

If the equality holds, then $gr_{\mathbb{M}}(M)$ is Cohen–Macaulay and hence M is Cohen–Macaulay.

Proof. We induct on r. Let $r = 1$ and $J = (a)$. We have

$$\frac{\lambda(M/M_1) + \lambda(M_1/JM)z}{(1-z)} = \lambda(M/M_1) + \lambda(M/aM) \sum_{j \geq 1} z^j$$

and $H_{\mathbb{M}}(0) = \lambda(M/M_1)$.

From the diagram

$$\begin{array}{ccc} M \supseteq M_n & \supseteq & M_{n+1} \\ \| & & \cup \\ M \supseteq aM & \supseteq & aM_n \end{array}$$

we get

$$\lambda(M/aM) + \lambda(aM/aM_n) = \lambda(M/M_n) + H_{\mathbb{M}}(n) + \lambda(M_{n+1}/aM_n).$$

On the other hand, from the exact sequence

$$0 \to (0 : a + M_n)/M_n \to M/M_n \xrightarrow{a} aM/aM_n \to 0$$

we get

$$\lambda(M/M_n) = \lambda(aM/aM_n) + \lambda((0 : a + M_n)/M_n).$$

It follows that for every $n \geq 1$

$$\lambda(M/aM) = H_{\mathbb{M}}(n) + \lambda(M_{n+1}/aM_n) + \lambda((0 : a + M_n)/M_n). \qquad (1.3)$$

This proves that $H_{\mathbb{M}}(n) \leq \lambda(M/aM)$ for every $n \geq 1$ and the first assertion of the theorem follows.

If we have equality above then, by (1.3), for every $n \geq 1$ we get

$$\lambda(M_{n+1}/aM_n) = \lambda((0 : a + M_n)/M_n) = 0.$$

This clearly implies that $M_{j+1} : a = M_j$ for every $j \geq 1$ so that, by Lemma 1.1, $a^* \in \mathfrak{q}/\mathfrak{q}^2$ is regular on $gr_{\mathbb{M}}(M)$ and $gr_{\mathbb{M}}(M)$ is Cohen–Macaulay.

Suppose $r \geq 2$, $J = (a_1, \ldots, a_r)$ and let us consider the good $\mathfrak{q}$-filtration $\mathbb{M}/a_1M$ on M/a_1M. We have $\dim M/a_1M = r - 1$ and we know that $a_2, \ldots, a_r$ is a maximal

M/a_1M-superficial sequence for q. By the inductive assumption and since $a_1M \subseteq M_1$, we get

$$P_{M/a_1M}(z) \leq \frac{\lambda((M/a_1M)/(a_1M+M_1/a_1M)) + \lambda((a_1M+M_1/a_1M)/K(M/a_1M))z}{(1-z)^{r-1}}$$

$$= \frac{\lambda(M/M_1) + \lambda(M_1/JM)z}{(1-z)^{r-1}}.$$

where we let $K := (a_2,\dots,a_r)$.

By using (1.2) and since the power series $\frac{1}{1-z}$ is positive, we get

$$P_M(z) \leq P^1_{M/a_1M}(z) = \frac{P_{M/a_1M}(z)}{1-z} \leq \frac{\lambda(M/M_1) + \lambda(M_1/JM)z}{(1-z)^r},$$

as wanted.

If we have equality, then

$$P_{M/a_1M}(z) = \frac{\lambda(M/M_1) + \lambda(M_1/JM)z}{(1-z)^{r-1}}$$

so that $gr_{M/a_1M}(M/a_1M)$ is Cohen–Macaulay and hence $gr_M(M)$ is Cohen–Macaulay as well by Sally's machine. In particular M is Cohen–Macaulay. □

The above result says that if the h-polynomial is $h_M(z) = \lambda(M/M_1) + \lambda(M_1/JM) z$, we may conclude that $gr_M(M)$ is Cohen–Macaulay even if we do not assume the Cohen–Macaulayness of M. The result cannot be extended to any *short h-polynomial* $h_M(z) = h_0 + h_1z$. For example if $A = k[[x,y]]/(x^2,xy,xz,y^3)$ and we consider the m-adic filtration, then $P_A(z) = \frac{1+2z}{1-z}$, but $gr_m(A) \simeq A$ is not Cohen–Macaulay.

In the classical case of the m-adic filtration on a local Cohen–Macaulay ring A, Elias and Valla in [26] proved that the h-polynomial of the form $h_m(z) = h_0 + h_1z + +h_2z^2$ forces $gr_m(A)$ to be Cohen–Macaulay. We cannot extend this result to general filtrations because this is not longer true even if we consider the q-adic filtration with q an m-primary ideal of a Cohen–Macaulay ring. The following example is due to Sally [92, Example 3.3].

Example 1.1. Let $A = k[[t^4,t^5,t^6,t^7]]$ and consider $q = (t^4,t^5,t^6)$. We have

$$P_q(z) = \frac{2+z+z^2}{(1-z)}$$

and $gr_q(A)$ is not Cohen–Macaulay because $a = t^4$ is a superficial (regular) element for q, but $q^2 : a \neq q$ (cf. Lemmas 1.1 and 1.3).

Chapter 2
Bounds for $e_0(\mathbb{M})$ and $e_1(\mathbb{M})$

In this chapter we prove lower and upper bounds for the first two coefficients of the Hilbert polynomial which we defined in Chap. 1. We recall that $e_0(\mathbb{M})$ depends only on q and M, but it does not depend on the good q-filtration $\mathbb{M}$. In contrast $e_1(\mathbb{M})$ does depend on the filtration $\mathbb{M}$. It is called by Vasconcelos *tracking number* for its *tag* position among the different filtrations having the same multiplicity. The coefficient $e_1(\mathbb{M})$ is also called the *Chern number* (see [111]).

In establishing the properties of $e_1(\mathbb{M})$, we will need an ad hoc treatment of the one-dimensional case. The results will be extended to higher dimensions via inductive arguments and via Proposition 1.2, which describes the behaviour of the Hilbert coefficients modulo superficial elements. In this way we prove and extend several classical bounds on $e_0(\mathbb{M})$ and $e_1(\mathbb{M})$ which we are going to describe. We start with Abhyankar–Valla formula, which gives a natural lower bound for the multiplicity of a Cohen–Macaulay filtered module M. The study of Cohen–Macaulay local rings of minimal multiplicity with respect to this bound, was carried out by J. Sally in [87]. This paper can be considered as the starting point of much of the recent research in this field. We extend here our interest to the non Cohen–Macaulay case taking advantage of the fact that the correction term we are going to introduce behaves well modulo superficial elements.

Concerning $e_1(\mathbb{M})$, we extend considerably the inequality

$$e_1(\mathfrak{m}) \geq e_0(\mathfrak{m}) - 1$$

proved by D.G. Northcott in [59]. Besides Northcott's inequality, Theorem 2.4 extends the corresponding inequality proved by Fillmore in [27] in the case of Cohen–Macaulay modules, by Guerrieri and Rossi in [38] for filtration of ideals and later by Puthenpurakal in [69] for q-adic filtrations of Cohen–Macaulay modules. When the module M is not necessarily Cohen–Macaulay, we present a new proof of a recent result by Goto and Nishida in [31]. In our general setting we will focus on an upper bound of $e_1(\mathbb{M})$ which was introduced and studied in the classical case by Huckaba and Marley in [53].

In the last section, we show that modules which are extremal with respect to the inequalities proved above have good associated graded modules and Hilbert functions of very specific shape. In some cases we shall see that extremal values

M.E. Rossi and G. Valla, *Hilbert Functions of Filtered Modules*, Lecture Notes of the Unione Matematica Italiana 9, DOI 10.1007/978-3-642-14240-6_2,

of the integer $e_1(\mathfrak{m})$ necessarily imply that the ring A is Cohen–Macaulay. These results can be considered as a confirmation of the general philosophy of the paper of W. Vasconcelos [111], where the Chern number is conjectured to be a measure of how far A is from being Cohen–Macaulay.

2.1 The Multiplicity and the First Hilbert Coefficient: Basic Facts

In [1], S. Abhyankar proved a nice lower bound for the multiplicity of a Cohen–Macaulay local ring $(A,\mathfrak{m})$. He found that

$$e_0(\mathfrak{m}) \geq b - r + 1$$

where r is the dimension of A and $b = H_A(1)$ is the embedding dimension of A.

G. Valla extended the formula to $\mathfrak{m}$-primary ideals in [106]. Guerrieri and Rossi in [38] showed that the result holds for ideal filtrations. In [69] Puthenpurakal proved the formula for Cohen–Macaulay modules and ideal filtrations by using the idealization of the module. Here we show that the original proof by Valla extends naturally to our general setting.

As before, we write the h-polynomial of M as

$$h_{\mathbb{M}}(z) = h_0(\mathbb{M}) + h_1(\mathbb{M})z + \cdots + h_s(\mathbb{M})z^s.$$

Hence, if $\dim(M) = r$, we have

$$h_0(\mathbb{M}) = \lambda(M/M_1) \quad h_1(\mathbb{M}) = \lambda(M_1/M_2) - r\lambda(M/M_1).$$

If a is an $\mathbb{M}$-superficial element for $\mathfrak{q}$, then

$$h_0(\mathbb{M}) = h_0(\mathbb{M}/aM) \text{ but in general } h_1(\mathbb{M}) \neq h_1(\mathbb{M}/aM).$$

If a is a regular element on M, then

$$h_1(\mathbb{M}/aM) = h_1(\mathbb{M}) + \lambda(M/M_1) - \lambda(M/M_2 : a) \geq h_1(\mathbb{M})$$

and $h_1(\mathbb{M}/aM) = h_1(\mathbb{M})$ if and only if $M_2 : a = M_1$.

In the classical case of the $\mathfrak{m}$-adic filtration on a local Cohen–Macaulay ring A, $h_1(\mathbb{M})$ is the embedding codimension of A; it is positive unless A is a regular ring. In particular $h_1(\mathbb{M}) = h_1(\mathbb{M}/aM)$. The inequality $h_1(\mathbb{M}) \leq h_1(\mathbb{M}/aM)$ can be strict. In Example 1.1 we have $h_1(\mathbb{M}) = 1 < h_1(\mathbb{M}/aM) = 2$.

This point makes a crucial difference between the $\mathfrak{m}$-adic filtration and more general filtrations and it justifies the new invariant $h(\mathbb{M}) := \lambda(M_1/JM + M_2)$ which will be introduced later (see (2.12)). Notice that if M is Cohen–Macaulay and J is the ideal generated by a maximal $\mathbb{M}$-superficial sequence such that $M_2 \cap JM = JM_1$, then

$$h_1(\mathbb{M}) = h_1(\mathbb{M}/JM).$$

The assumption $M_2 \cap JM = JM_1$ is verified if $\mathbb{M}$ is the q-adic filtration on a Cohen–Macaulay local ring A and q is integrally closed (see [54] and [58]).

Proposition 2.1. *Let M be a module of dimension $r \geq 1$ and let $J = (a_1, \ldots, a_r)$ be the ideal generated by an $\mathbb{M}$-superficial sequence for the $\mathfrak{m}$-primary ideal q. If $L := (a_1, \ldots, a_{r-1})$ and $I \supseteq$ q is an ideal of A, then*

$$e_0(\mathbb{M}) = h_0(\mathbb{M}) + h_1(\mathbb{M}) + r\lambda(M/M_1) - \lambda(JM/JM_1)$$
$$+ \lambda(M_2/JM_1) - \lambda(LM : a_r/LM)$$
$$= h_0(\mathbb{M}) + \lambda(M_1/IM_1) - \lambda(JM/IJM)$$
$$+ \lambda(IM_1/IJM) - \lambda(LM : a_r/LM). \qquad (2.1)$$

Proof. By using Proposition 1.2, we get

$$e_0(\mathbb{M}) = e_0(\mathbb{M}/LM) = e_0(\mathbb{M}/JM) - \lambda(LM : a_r/LM)$$
$$= \lambda(M/JM) - \lambda(LM : a_r/LM)$$

The conclusion easily follows by using the following diagrams

$$
\begin{array}{ccc}
M \supset M_1 \supset JM & \qquad & M \supset M_1 \supset IM_1 \\
\cup \qquad \cup & & \cup \qquad \cup \\
M_2 \supset JM_1 & & JM \supset IJM \qquad \qquad \square
\end{array}
$$

If M is Cohen–Macaulay, since q is $\mathfrak{m}$-primary, then $\mathrm{depth}_q(M) = \mathrm{depth}_\mathfrak{m}(M) = r$ and the elements $a_1, \ldots, a_r$ form a regular sequence on M. Since $J = (a_1, \cdots, a_r)$ is generated by a regular sequence and $JM \subseteq M_1$, we get $r\lambda(M/M_1) = \lambda(JM/JM_1)$ and $\lambda(JM/IJM) = r\lambda(M/IM)$. Moreover $\lambda(LM : a_r/LM) = 0$.

Thus, as a consequence of the above proposition, we get the following result.

Corollary 2.1. *Let M be a Cohen–Macaulay module of dimension $r \geq 1$, J the ideal generated by a maximal $\mathbb{M}$-superficial sequence for q and I an ideal containing q. Then*

$$e_0(\mathbb{M}) = h_0(\mathbb{M}) + h_1(\mathbb{M}) + \lambda(M_2/JM_1)$$
$$= \lambda(M/M_1) - r\lambda(M/IM) + \lambda(M_1/IM_1) + \lambda(IM_1/JIM).$$

The above result shows in particular that, if M is Cohen–Macaulay, then $\lambda(M_2/JM_1)$ does not depend on J. The first formula was proved by Abhyankar for the $\mathfrak{m}$-adic filtration and by Valla for an $\mathfrak{m}$-primary ideal q. The second formula is due to Goto who proved it in the case of the q-adic filtration and $I = \mathfrak{m}$.

Following the notation of the pioneering work of J. Sally, it is natural to give the following definition.

Let M be Cohen–Macaulay and $\mathbb{M}$ a good q-filtration. We say that the filtration $\mathbb{M}$ has *minimal multiplicity* if

$$e_0(\mathbb{M}) = h_0(\mathbb{M}) + h_1(\mathbb{M})$$

or equivalently if $M_2 = JM_1$.

Following [30], we say that the filtration $\mathbb{M}$ has *Goto minimal multiplicity* with respect to the ideal I if

$$e_0(\mathbb{M}) = \lambda(M/M_1) - r\lambda(M/IM) + \lambda(M_1/IM_1)$$

or equivalently $IM_1 = JIM$.

Let us compare these two definitions. Given a good q-filtration $\mathbb{M}$ on the module M and an ideal $I \supseteq \mathfrak{q}$, we define a new filtration $\mathbb{M}^I$ on M as follows:

$$\mathbb{M}^I: \quad M \supseteq IM \supseteq IM_1 \supseteq \cdots \supseteq IM_n \cdots \supseteq \ldots \tag{2.2}$$

It is clear that $\mathbb{M}^I$ is a good q-filtration on M so that $e_0(\mathbb{M}) = e_0(\mathbb{M}^I)$.

Proposition 2.2. $\mathbb{M}$ *has Goto minimal multiplicity with respect to I if and only if* $\mathbb{M}^I$ *has minimal multiplicity.*

Proof. $\mathbb{M}^I$ has minimal multiplicity if and only if

$$e_0(\mathbb{M}^I) = h_0(\mathbb{M}^I) + h_1(\mathbb{M}^I)$$

This means

$$e_0(\mathbb{M}) = \lambda(M/IM) + \lambda(IM/IM_1) - r\lambda(M/IM).$$

Since we have a diagram

$$\begin{array}{ccc} M & \supset & IM \\ \cup & & \cup \\ M_1 & \supset & IM_1 \end{array}$$

we get that $\mathbb{M}^I$ has minimal multiplicity if and only if

$$e_0(\mathbb{M}) = \lambda(M/M_1) + \lambda(M_1/IM_1) - r\lambda(M/IM).$$

This is exactly the condition for $\mathbb{M}$ to have Goto minimal multiplicity.   □

If $\mathfrak{q}$ is an m-primary ideal of a local Cohen–Macaulay ring $(A, \mathfrak{m})$ of dimension r, and $\mathbb{M}$ is the q-adic filtration, we get that $e_0(\mathfrak{q})$ is minimal if and only if $\mathfrak{q}^2 = J\mathfrak{q}$. It is clear that this implies that $gr_\mathfrak{q}(A)$ is Cohen–Macaulay by Valabrega–Valla.

On the other hand, by its definition, the q-adic filtration has Goto minimal multiplicity with respect to m if and only if $\mathfrak{q}\mathfrak{m} = J\mathfrak{m}$.

The two notions coincide if $q = m$. We remark that if q is integrally closed and if the q-adic filtration has Goto minimal multiplicity, then it has minimal multiplicity. In fact the condition $qm = Jm$ implies $q^2 \subseteq J$, hence $q^2 = Jq$ since the integrality guarantees $q^2 \cap J = Jq$ (see [54] and [58]). The converse is no longer true.

In general the condition $qm = Jm$ seems far from imposing any restriction on the Hilbert function of q. Notice that it does not imply that $gr_q(A)$ is Cohen–Macaulay, nor even that $gr_q(A)$ is Buchsbaum (see [31, Theorem 3.1]).

We will study the Hilbert function in the case of minimal multiplicity in Theorem 2.9 and Corollary 2.6.

We introduce now the notion of almost minimal multiplicity. Given a good q-filtration $\mathbb{M}$ of a Cohen–Macaulay module M, we say that $\mathbb{M}$ has *almost minimal multiplicity* if

$$e_0(\mathbb{M}) = h_0(\mathbb{M}) + h_1(\mathbb{M}) + 1,$$

or equivalently $\lambda(M_2/JM_1) = 1$.

Analogously we will say that $\mathbb{M}$ has *Goto almost minimal multiplicity* if and only if $\mathbb{M}^I$ has almost minimal multiplicity, equivalently $\lambda(IM_1/JM_1) = 1$ for every J generated by a maximal $\mathbb{M}$-superficial sequence for q.

We will investigate the Hilbert function of $\mathbb{M}$ when it has almost minimal multiplicity. The problem is far more difficult; it amounts to the Sally's conjecture (see [91]) which was open for several years and finally solved by Wang and independently by Rossi and Valla (see [80] and [115]).

We notice that the concept of almost minimal multiplicity introduced by Jayanthan and Verma (see [43]), is equivalent to saying that the filtration $\mathbb{M}^I$ has almost minimal multiplicity.

That's all for the moment, as far as we are concerned with bounds for e_0. Instead, let us come to the first Hilbert coefficient e_1.

When q is an m-primary ideal of the Cohen–Macaulay local ring (A, m), $M = A$ and $\mathbb{M}$ the q-adic filtration, in order to prove that $e_1(q) \geq 0$, Northcott proved a basic lower bound for e_1, namely $e_1(q) \geq e_0(q) - \lambda(A/q) \geq 0$ (see [64]). Fillmore extended it to Cohen–Macaulay modules in [27] (see also [69]) and later Huneke (see [54]) and Ooishi (see [66]) proved that $e_1(q) = e_0(q) - \lambda(A/q)$ if and only if $q^2 = Jq$, where J is the ideal generated by a maximal superficial sequence for q. When this is the case, by the Valabrega–Valla criterion, the associated graded ring is Cohen–Macaulay and the Hilbert function is easily described. This result has been extended to ideal filtrations of Cohen–Macaulay rings by Guerrieri and Rossi in [38]. Recently Goto and Nishida in [31] generalized the inequality, with suitable correction terms, to any local ring not necessarily Cohen–Macaulay and they studied the equality in the Buchsbaum case.

The result of Northcott was improved by Elias and Valla in [26] where, for the maximal ideal of a Cohen–Macaulay local ring (A, m), one can find a proof of the inequality

$$e_1(m) \geq 2e_0(m) - h - 2,$$

where $h = \mu(\mathfrak{m}) - \dim(A)$ is the embedding codimension of A. Since by Abhyankar

$$2e_0(\mathfrak{m}) - h - 2 \geq e_0(\mathfrak{m}) - 1,$$

this is also an extension of Northcott's inequality. Further, it has been proved that, if equality holds, then the associated graded ring is Cohen–Macaulay and the Hilbert function is determined. This result was extended to Hilbert filtrations of ideals by Guerrieri and Rossi in [38] and rediscovered, only for the q-adic filtration associated to an $\mathfrak{m}$-primary ideal q, by Corso, Polini and Vasconcelos in [15, (2.9)]. When equality holds, they need an extra assumption on the Sally module, in order to get a Cohen–Macaulay associated graded ring. We note that this extra assumption is not essential as already proved in [38, Theorem 2.2]. Recently Corso in [12] was able to remove the Cohen–Macaulayness assumption. Here we extend and improve at the same time all these results in our general setting, by using a very simple inductive argument.

Other notable bounds will be presented.

2.2 The One-Dimensional Case

In establishing the properties of the Hilbert coefficients of a filtered module M, it will be convenient to use induction on the dimension of the module. To start the induction, we need an ad hoc treatment of the one-dimensional case. One-dimensional Cohen–Macaulay rings have been extensively studied in the classical case by Matlis in [62]. We present here a general approach for filtered modules which are not necessarily Cohen–Macaulay.

Given a module M *of dimension one* and a good q-filtration $\mathbb{M} = \{M_j\}_{j \geq 0}$ of M, we know that for large n we have $e_0(\mathbb{M}) = H_{\mathbb{M}}(n)$, so that we define for every $j \geq 0$ the integers

$$u_j(\mathbb{M}) := e_0(\mathbb{M}) - H_{\mathbb{M}}(j). \tag{2.3}$$

Lemma 2.1. *Let M be a module of dimension one. If a is an $\mathbb{M}$-superficial element for q, then for every $j \geq 0$ we have*

$$u_j(\mathbb{M}) = \lambda(M_{j+1}/aM_j) - \lambda(0 :_{M_j} a).$$

Proof. By Proposition 1.2(3), we have

$$e_0(\mathbb{M}) = e_0(\mathbb{M}/aM) - \lambda(0 :_M a) = \lambda(M/aM) - \lambda(0 :_M a)$$
$$= \lambda(M/aM_j) - \lambda(aM/aM_j) - \lambda(0 :_M a).$$

By using the following exact sequence

$$0 \to (0 :_M a)/(0 :_{M_j} a) \to M/M_j \to aM/aM_j \to 0$$

we get

$$e_0(\mathbb{M}) = \lambda(M/aM_j) - \lambda(M/M_j) + \lambda((0 :_M a)/(0 :_{M_j} a)) - \lambda(0 :_M a)$$

and finally

$$\begin{aligned}
u_j(\mathbb{M}) &= e_0(\mathbb{M}) - \lambda(M_j/M_{j+1}) \\
&= \lambda(M/aM_j) - \lambda(M/M_j) - \lambda(0 :_{M_j} a) - \lambda(M_j/M_{j+1}) \\
&= \lambda(M_{j+1}/aM_j) - \lambda(0 :_{M_j} a).
\end{aligned}$$

$\square$

It follows that, when M is one-dimensional and Cohen–Macaulay, then $u_j(\mathbb{M}) = \lambda(M_{j+1}/aM_j)$ is non negative and we have, for every $j \geq 0$,

$$H_{\mathbb{M}}(j) = e_0(\mathbb{M}) - \lambda(M_{j+1}/aM_j) \leq e_0(\mathbb{M}).$$

It will be useful to write down the Hilbert coefficients in terms of the integers $u_j(\mathbb{M})$.

Lemma 2.2. *Let M be a module of dimension one. Then for every $j \geq 1$ we have*

$$e_j(\mathbb{M}) = \sum_{k \geq j-1} \binom{k}{j-1} u_k(\mathbb{M}).$$

Proof. We have

$$P_{\mathbb{M}}(z) = \frac{h_{\mathbb{M}}(z)}{1-z} = \sum_{j \geq 0} H_{\mathbb{M}}(j) z^j.$$

Hence, if we write $h_{\mathbb{M}}(z) = h_0(\mathbb{M}) + h_1(\mathbb{M})z + \cdots + h_s(\mathbb{M})z^s$, then we get for every $k \geq 1$

$$h_k(\mathbb{M}) = H_{\mathbb{M}}(k) - H_{\mathbb{M}}(k-1) = u_{k-1}(\mathbb{M}) - u_k(\mathbb{M}).$$

Hence we can compute the Hilbert series of M

$$P_{\mathbb{M}}(z) = \frac{e_0(\mathbb{M}) - u_0(\mathbb{M}) + \sum_{k \geq 1}(u_{k-1}(\mathbb{M}) - u_k(\mathbb{M}))z^k}{(1-z)} \qquad (2.4)$$

Finally

$$\begin{aligned}
e_j(\mathbb{M}) &= \frac{h_{\mathbb{M}}^{(j)}(1)}{j!} = \sum_{k \geq j} \binom{k}{j} h_k(\mathbb{M}) = \sum_{k \geq j} \binom{k}{j}(u_{k-1}(\mathbb{M}) - u_k(\mathbb{M})) \\
&= \sum_{k \geq j-1} \binom{k}{j-1} u_k(\mathbb{M}).
\end{aligned}$$

$\square$

If we apply the above Lemma when M is Cohen–Macaulay, and using the fact that the integers $u_k(\mathbb{M})$ are non negative, we get

$$e_1(\mathbb{M}) = \sum_{k \geq 0} u_k(\mathbb{M}) \geq u_0(\mathbb{M}) + u_1(\mathbb{M}) \geq u_0(\mathbb{M}).$$

Since we have $u_0(\mathbb{M}) = e_0(\mathbb{M}) - \lambda(M/M_1)$, and $u_0(\mathbb{M}) + u_1(\mathbb{M}) = 2e_0(\mathbb{M}) - \lambda(M/M_2)$, we trivially get

$$e_1(\mathbb{M}) \geq e_0(\mathbb{M}) - \lambda(M/M_1),$$

and

$$e_1(\mathbb{M}) \geq 2e_0(\mathbb{M}) - \lambda(M/M_2),$$

which are the bounds proved by Northcott and by Elias–Valla, in the one-dimensional Cohen–Macaulay case.

If we do not assume that M is Cohen–Macaulay, then the integers $u_i(\mathbb{M})$ can be negative, so that the equality $e_1(\mathbb{M}) = \sum_{k \geq 0} u_k(\mathbb{M})$ no longer implies $e_1(\mathbb{M}) \geq u_0(\mathbb{M})$. For instance, if we consider the local ring $A = k[[X,Y]]/(X^2,XY)$ endowed with the m-adic filtration, we have $e_0 = 1$, $e_1 = -1$ and $u_0 = 0$.

Hence we need to change this formula by introducing some correction terms which vanish in the Cohen–Macaulay case.

Given a good q-filtration $\mathbb{M} = \{M_j\}_{j \geq 0}$ of the r-dimensional module M, let $a_1, \ldots, a_r$ be an $\mathbb{M}$-superficial sequence for q; further let $J := (a_1, \ldots, a_r)$ and

$$\mathbb{N} := \{J^j M\}$$

be the J-adic filtration on M which is clearly J-good. It is not difficult to prove that also the original filtration $\mathbb{M}$ is J-good and this implies that $e_0(\mathbb{M}) = e_0(\mathbb{N})$.

When M is Cohen–Macaulay, the elements $a_1, \ldots, a_r$ form a regular sequence on M and, as a consequence, one can prove

$$J^i M / J^{i+1} M \simeq (M/JM)^{\binom{r+i-1}{i}}.$$

This implies that the Hilbert Series of $\mathbb{N}$ is $P_{\mathbb{N}}(z) = \frac{\lambda(M/JM)}{(1-z)^r}$ and thus $e_i(\mathbb{N}) = 0$ for every $i \geq 1$. This proves that these integers give a good measure of how M differs from being Cohen–Macaulay.

We prove that $e_1(\mathbb{N}) \leq 0$ in the one-dimensional case by relating the integer $e_1(\mathbb{N})$ to the 0-th local cohomology module of M.

The following Lemma is a first confirmation of a conjecture stated by Vasconcelos in [111] concerning the negativity of $e_1(J)$ in the higher dimensional case. Surprising results concerning this problem have been recently proved in [29] and in [28].

In the following we denote by W the 0-th local cohomology module $H^0_{\mathfrak{m}}(M)$ of M with respect to $\mathfrak{m}$. We know that $H^0_{\mathfrak{m}}(M) := \cup_{j \geq 0}(0 :_M \mathfrak{m}^j) = 0 :_M \mathfrak{m}^t$ for every $t \gg 0$.

In the one-dimensional case we have the following nice formula (see [31, Lemma 2.4]).

Lemma 2.3. *Let M be a finitely generated A-module of dimension one and let a be a parameter for M. Then for every $t \gg 0$ we have $W = 0 :_M a^t$ and, if we denote by $\mathbb{N}$ the (a)-adic filtration on M, then*

$$\lambda(W) = -e_1(\mathbb{N}).$$

Proof. Since there is an integer j such that $\mathfrak{m}^j M \subseteq aM = ((a) + 0 : M)M$, the ideal $(a) + 0 : M$ is $\mathfrak{m}$-primary and therefore $\mathfrak{m}^s \subseteq (a) + 0 : M$ for some s; this implies

$$\mathfrak{m}^{ts} \subseteq (a)^t + 0 : M$$

for every t. On the other hand, $W = 0 :_M \mathfrak{m}^t$ for every integer $t \gg 0$, so that

$$W = 0 :_M \mathfrak{m}^t \subseteq 0 :_M a^t \subseteq 0 :_M \mathfrak{m}^{ts} = W.$$

We denote by $\mathbb{N}^n$ the (a^n)-adic filtration on M. Now for $n \gg 0$, it is easy to see that $ne_0(\mathbb{N}) = e_0(\mathbb{N}^n) = \lambda(M/a^n M) - \lambda(0 :_M a^n)$ and the result follows because $\lambda(M/a^n M) = ne_0(\mathbb{N}) - e_1(\mathbb{N})$. $\qquad\square$

Given a good q-filtration of the module M (of any dimension), we consider now the corresponding filtration of the saturated module $M^{sat} := M/W$. This is the filtration
$$\mathbb{M}^{sat} := \mathbb{M}/W = \{M_n + W/W\}_{n \geq 0}.$$

Since W has finite length and $\cap_{i \geq 0} M_i = \{0\}$, we have $M_i \cap W = \{0\}$ for every $i \gg 0$. This implies $p_{\mathbb{M}}(X) = p_{\mathbb{M}^{sat}}(X)$.

Further, it is clear that for every $j \geq 0$ we have an exact sequence

$$0 \to W/(M_{j+1} \cap W) \to M/M_{j+1} \to M/(M_{j+1} + W) \to 0$$

so that for every $j \gg 0$ we have

$$\lambda(M/M_{j+1}) = \lambda[M/(M_{j+1} + W)] + \lambda(W)$$

which implies

$$p^1_{\mathbb{M}}(X) = p^1_{\mathbb{M}^{sat}}(X) + \lambda(W). \tag{2.5}$$

This proves the following result:

Proposition 2.3. *Let M be a module of dimension r. Denote $W := H^0_{\mathfrak{m}}(M)$ and $\mathbb{M}^{sat} := \mathbb{M}/W$. Then*

$$e_i(\mathbb{M}) = e_i(\mathbb{M}^{sat}) \; 0 \leq i \leq r-1, \qquad e_r(\mathbb{M}) = e_r(\mathbb{M}^{sat}) + (-1)^d \lambda(W).$$

We remark that, if $\dim(M) \geq 1$, the module M/W always has positive depth. This is the reason why, sometimes, we move our attention from the module M to the module M/W. This will be the strategy of the proof of the next proposition which gives, in the one-dimensional case, the promised upper bound for e_1.

Proposition 2.4. *Let* $\mathbb{M} = \{M_j\}_{j \geq 0}$ *be a good* q-*filtration of a module M of dimension one. If a is an $\mathbb{M}$-superficial element for* q *and* $\mathbb{N}$ *the* (a)-*adic filtration on M, then*

$$e_1(\mathbb{M}) - e_1(\mathbb{N}) \leq \sum_{j \geq 0} \lambda(M_{j+1}/aM_j).$$

If $W \subseteq M_1$ *and equality holds above, then M is Cohen–Macaulay.*

Proof. By Propositions 2.3 and 2.3 we have

$$e_1(\mathbb{M}) = e_1(\mathbb{M}^{sat}) - \lambda(W) = e_1(\mathbb{M}^{sat}) + e_1(\mathbb{N})$$

so that we need to prove that $e_1(\mathbb{M}^{sat}) \leq \sum_{j \geq 0} \lambda(M_{j+1}/aM_j)$.

Now M/W is Cohen–Macaulay and a is regular on M/W, hence by Lemmas 2.2 and 2.1, we get

$$e_1(\mathbb{M}^{sat}) = \sum_{j \geq 0} u_j(\mathbb{M}^{sat}) = \sum_{j \geq 0} \lambda(M_{j+1}^{sat}/aM_j^{sat})$$

$$= \sum_{j \geq 0} \lambda \left[\frac{M_{j+1} + W}{aM_j + W} \right] = \sum_{j \geq 0} \lambda \left[\frac{M_{j+1}}{aM_j + M_{j+1} \cap W} \right]$$

$$\leq \sum_{j \geq 0} \lambda(M_{j+1}/aM_j).$$

The first assertion follows. In particular equality holds if and only if we get $M_{j+1} \cap W \subseteq aM_j$ for every $j \geq 0$. Let as assume $W \subseteq M_1$ and equality above; then we have $W = W \cap M_1 \subseteq aM$. Now recall that $W = 0 :_M a^t$ for $t \gg 0$, hence if $c \in W$ then $c = am$ with $a^t c = a^{t+1} m = 0$. This implies $m \in 0 :_M a^{t+1} = W$ so that $W \subseteq aW$ and, by Nakayama, $W = 0$. $\square$

We turn out to describing lower bounds on the first Hilbert coefficient thus extending the classical result proved by Northcott.

Proposition 2.5. *Let* $\mathbb{M} = \{M_j\}_{j \geq 0}$ *be a good* q-*filtration of a module M of dimension one. If a is an $\mathbb{M}$-superficial element for* q *and $s \geq 1$ a given integer, then for every $n \gg 0$ we have*

$$e_1(\mathbb{M}) - e_1(\mathbb{N}) = se_0(\mathbb{M}) - \lambda(M/M_s) + \lambda(M_s + W/M_s) + \lambda(M_n/a^{n-s}M_s)$$

$$= \sum_{j=0}^{s-1} u_j(\mathbb{M}) + \lambda(M_s + W/M_s) + \lambda(M_n/a^{n-s}M_s).$$

Proof. We have for every $n \gg 0$ the following equalities:

$$\lambda(M/M_n) = p_{\mathbb{M}}^1(n-1) = e_0(\mathbb{M})n - e_1(\mathbb{M})$$

$$\lambda(M/a^{n-s}M) = p_{\mathbb{N}}^1(n-s-1) = e_0(\mathbb{N})(n-s) - e_1(\mathbb{N}).$$

Since $e_0(\mathbb{M}) = e_0(\mathbb{N})$, we get

$$e_1(\mathbb{M}) - e_1(\mathbb{N}) = s\,e_0(\mathbb{M}) - \lambda(M/M_n) + \lambda(M/a^{n-s}M).$$

From the diagram

$$
\begin{array}{ccc}
M & \supset & M_n \\
\cup & & \cup \\
a^{n-s}M & \supset & a^{n-s}M_s
\end{array}
$$

we get

$$e_1(\mathbb{M}) - e_1(\mathbb{N}) = s\,e_0(\mathbb{M}) + \lambda(M_n/a^{n-s}M_s) - \lambda(a^{n-s}M/a^{n-s}M_s).$$

By using the exact sequence

$$0 \to (M_s + 0 :_M a^{n-s}/M_s) \longrightarrow M/M_s \xrightarrow{a^{n-s}} a^{n-s}M/a^{n-s}M_s \longrightarrow 0$$

and the equality $0 :_M a^t = W$ for $t \gg 0$, we get the conclusion. $\qquad\square$

Corollary 2.2. *With the same notation as in the above Proposition, if*

$$e_1(\mathbb{M}) - e_1(\mathbb{N}) = s\,e_0(\mathbb{M}) - \lambda(M/M_s) + \lambda(M_s + W/M_s),$$

then $M_{s+1} \subseteq aM_s + W$.

Proof. We simply notice that we have an injective map

$$(M_{s+1} + W)/(aM_s + W) \xrightarrow{a^{n-s-1}} M_n/a^{n-s}M_s.$$

$\qquad\square$

The converse does not hold, as the following example shows. Let $A = k[[t^3, t^4, t^5]]$ and let us consider the following m-filtration $\mathbb{M}$ on A :

$$M = A, \quad M_1 = \mathfrak{m}, \quad M_2 = \mathfrak{m}^2, \quad M_3 = \mathfrak{m}^2, \quad M_j = \mathfrak{m}^{j-1}$$

for $j \geq 4$. It is clear that t^3 is an $\mathbb{M}$-superficial element for m and A is Cohen–Macaulay so that $W = 0$. We have

$$P_{\mathbb{M}}(z) = \frac{1 + 2z - 3z^2 + 3z^3}{1 - z}, \quad P_{\mathbb{N}}(z) = \frac{3}{1 - z}$$

so that $e_0(\mathbb{M}) = 3$, $e_1(\mathbb{M}) = 5$, $e_1(\mathbb{N}) = 0$ and the equality $e_1(\mathbb{M}) - e_1(\mathbb{N}) = e_0(\mathbb{M}) - \lambda(M/M_1)$ does not hold even if $M_2 = t^3 M_1$.

The following result was proved in [31, Lemma 2.1] in the case $M = A$ and $s = 1$.

Corollary 2.3. *Let* $\mathbb{M} = \{\mathfrak{q}^j M\}_{j \geq 0}$ *be the* $\mathfrak{q}$*-adic filtration on* M *of dimension one. Let* $a \in \mathfrak{q}$ *be an* $\mathbb{M}$*-superficial element for* $\mathfrak{q}$ *and* $s \geq 1$ *a given integer. Then*

$$e_1(\mathbb{M}) - e_1(\mathbb{N}) = s\,e_0(\mathbb{M}) - \lambda(M/M_s)$$

if and only if $M_{s+1} \subseteq aM_s + W$ *and* $W \subseteq M_s$.

Proof. By using Corollary 2.2 and Proposition 2.5, it is enough to prove that $M_{s+1} \subseteq aM_s + W$ implies $M_n = a^{n-s} M_s$ for $n \gg 0$.

We have $M_{s+1} \subseteq aM_s + W$ and by multiplication by $\mathfrak{q}^{n-s}$ the result follows since $M_{n+1} \subseteq aM_n + \mathfrak{q}^{n-s}W = aM_n$ for every $n \gg 0$. $\square$

The above corollary holds true even if we consider the more general filtration later defined in (2.18).

Under the assumption $\dim M = 1$, we recover Theorem 1.3 in [31] and we give a positive answer to a question raised by Corso in [12].

Theorem 2.1. *Let* $\mathbb{M} = \{\mathfrak{m}^j M\}_{j \geq 0}$ *be the* $\mathfrak{m}$*-adic filtration of a Buchsbaum module* M *of dimension one. Assume either*

(i) $e_1(\mathbb{M}) - e_1(\mathbb{N}) = e_0(\mathbb{M}) - \lambda(M/M_1)$

or

(ii) $e_1(\mathbb{M}) - e_1(\mathbb{N}) = 2\,e_0(\mathbb{M}) - \lambda(M/M_2)$.

Then $gr_{\mathbb{M}}(M)$ *is Buchsbaum.*

Proof. By the above corollary, if either (i) or (ii) holds, then $M_{n+1} \subseteq aM_n + W$ for every $n \geq 2$. Then by Valabrega–Valla criterion applied to $\mathbb{M}^{sat} = \mathbb{M}/W$, it follows that $gr_{\mathbb{M}^{sat}}(M^{sat})$ is Cohen–Macaulay. Denote by Q the graded maximal ideal of $gr_{\mathfrak{m}}(A)$. By using the fact that $gr_{\mathbb{M}^{sat}}(M^{sat})$ is Cohen–Macaulay and $\mathfrak{m}W = 0$, it easy to see that $Q\,H^0_Q(gr_{\mathbb{M}}(M)) = 0$ and the result follows by [99, Proposition 2.12]. $\square$

For further applications, we need to consider another filtration related to a superficial sequence. Given a good $\mathfrak{q}$-filtration $\mathbb{M} = \{M_j\}_{j \geq 0}$ of the r-dimensional module M, let $a_1, \ldots, a_r$ be an $\mathbb{M}$-superficial sequence for $\mathfrak{q}$ and let $J := (a_1, \ldots, a_r)$. We define the following filtration

$$\mathbb{E}: \quad M \supseteq M_1 \supseteq JM_1 \supseteq J^2 M_1 \supseteq \cdots \supseteq J^j M_1 \supseteq J^{j+1}M_1 \supseteq \cdots$$

When $\mathbb{M}$ is the J-adic filtration on $M = A$, then we will write $e_i(J)$.

This filtration is J-good and we want to compare it with $\mathbb{N}$, the J-adic filtration on M. We need to remark that we have

$$e_0(\mathbb{E}) = e_0(\mathbb{M}) = e_0(\mathbb{N}).$$

Proposition 2.6. *Let* $\mathbb{M} = \{M_j\}_{j \geq 0}$ *be a good* $\mathfrak{q}$*-filtration on M of dimension one. If a is an* $\mathbb{M}$*-superficial element for* $\mathfrak{q}$*, then we have*

$$e_1(\mathbb{E}) - e_1(\mathbb{N}) = e_0(\mathbb{M}) - h_0(\mathbb{M}) + \lambda(M_1 + W/M_1).$$

In particular $e_1(\mathbb{E}) - e_1(\mathbb{N}) = e_0(\mathbb{M}) - h_0(\mathbb{M})$ *if and only if* $M_1 \supseteq W$.

Proof. As in the above proposition, we have for every $n \gg 0$ the following equalities:

$$\lambda(M/a^n M_1) = p_{\mathbb{E}}^1(n) = e_0(\mathbb{M})(n+1) - e_1(\mathbb{E})$$

$$\lambda(M/a^n M) = p_{\mathbb{N}}^1(n-1) = e_0(\mathbb{M})n - e_1(\mathbb{N}).$$

We get

$$\begin{aligned}
e_1(\mathbb{E}) - e_1(\mathbb{N}) &= e_0(\mathbb{M}) - \lambda(M/a^n M_1) + \lambda(M/a^n M) \\
&= e_0(\mathbb{M}) - \lambda(a^n M/a^n M_1) \\
&= e_0(\mathbb{M}) - \lambda(M/M_1) + \lambda(M_1 + W/M_1)
\end{aligned}$$

where the last equality follows from the exact sequence

$$0 \to (M_1 + W)/M_1 \to M/M_1 \xrightarrow{a^n} a^n M/a^n M_1 \to 0.$$

$\qquad\square$

We conclude this section concerning the one-dimensional case, with a notable extension of a bound on e_1 proved in the classical case by D. Kirby and extended to an m-primary ideal by M.E. Rossi, G. Valla and W. Vasconcelos. We assume M is a Cohen–Macaulay module of dimension one. We recall that the integers u_j are non negative because we have

$$u_j(\mathbb{M}) = \lambda(M_{j+1}/a M_j).$$

In particular for every $j \geq 0$ we have

$$H_{\mathbb{M}}(j) = e_0(\mathbb{M}) - \lambda(M_{j+1}/a M_j) \leq e_0(\mathbb{M})$$

where a is an $\mathbb{M}$-superficial element for $\mathfrak{q}$. If $\mathbb{M}$ is the $\mathfrak{q}$-adic filtration of a Cohen–Macaulay module M, then it is easy to see that

$$u_j(\mathbb{M}) = 0 \implies u_t(\mathbb{M}) = 0 \ \forall t \geq j.$$

In this case we define

$$s(\mathbb{M}) := \min\{j : H_{\mathbb{M}}(j) = e_0(\mathbb{M})\} = \min\{j : M_{j+1} = a M_j\}. \tag{2.6}$$

This integer coincides with the reduction number of $\mathbb{M}$ and from the above equality it is clear that it does not depend on (a). Moreover (2.4) shows that if M is a one-dimensional Cohen–Macaulay module, then $s(\mathbb{M})$ is also the degree of the h-polynomial of $\mathbb{M}$.

If A is a Cohen–Macaulay local ring of dimension one and we consider the classical m-adic filtration on A, Sally proved that

$$s(\mathfrak{m}) \le e_0(\mathfrak{m}) - 1$$

(see [94] and [92]). This result easily follows by a lower bound on the minimal number of generators proved by Herzog and Waldi [47, Theorem 2.1]. Herzog and Waldi's result can be generalized to modules in the case of the q-adic filtration on M where q is an m-primary ideal. The proof is a straightforward adaptation of the classical case; we describe it here for completeness. In the following $\mu(N)$ denotes the minimal number of generators of an A-module N on a local ring $(A, \mathfrak{m}, k)$, that is $\dim_k N/\mathfrak{m}N$. We may assume $s(\mathbb{M}) \ge 1$. In fact we have $s(\mathbb{M}) = 0$ if and only if $qM = aM$ where a is a regular $\mathbb{M}$-superficial element for q, hence $M_n \simeq aM$ for every $n \ge 1$. If $M = A$ this means that the local ring is regular.

Theorem 2.2. *Let $\mathbb{M}$ be the q-adic filtration of a Cohen–Macaulay module M of dimension one. Then*

(1) $\mu(q^n M) > n$ for every $n \le s(\mathbb{M})$.

(2) $\mu(q^n M) = \mu(q^{n+1} M)$ for every $n \ge s(\mathbb{M})$.

Proof. We prove (1). Let a be an $\mathbb{M}$-superficial element for q, since $s = s(\mathbb{M}) \ge 1$ and $q^s M \ne a q^{s-1} M$, then there exist $x_1, \ldots, x_s \in q$ and $m \in M$ such that

$$x_1 \cdots x_s m \in q^s M, \quad \text{but} \quad x_1 \cdots x_s m \notin a q^{s-1} M + \mathfrak{m} q^s M.$$

For every $i = 0, \ldots, s$ we consider

$$y_i := a^i x_{i+1} \cdots x_s m$$

and we claim that $\{y_0 = x_1 \cdots x_s m, \ y_1 = a x_2 \cdots x_s m, \ldots\ldots, y_s = a^s m\}$ is part of a minimal system of generators of $q^s M$. Assume $r_0 y_0 + r_1 y_1 + \cdots + r_s y_s \in \mathfrak{m} q^s M$ with $r_i \in A$ and we conclude $r_i \in \mathfrak{m}$ by arguing step by step on $i = 0, \ldots, s$. First $r_0 \in \mathfrak{m}$ otherwise $y_0 \in a q^{s-1} M + \mathfrak{m} q^s M$, Assume $i > 0$ and $r_0, \ldots, r_{i-1} \in \mathfrak{m}$, hence $r_i y_i + r_{i+1} y_{i+1} + \cdots + r_s y_s \in \mathfrak{m} q^s M$. Multiplying the sum with $x_1 \cdots x_i$ we obtain

$$a^i (r_i y_0 + a^{-i} r_{i+1} y_{i+1} x_1 \cdots x_i + \cdots + a^{-i} r_s y_s x_1 \cdots x_i) \in \mathfrak{m} q^{s+i} M.$$

Since $q^{s+i} M = a^i q^s M$ and a is regular on M, we get

$$r_i y_0 + a^{-i} r_{i+1} y_{i+1} x_1 \cdots x_i + \cdots + a^{-i} r_s y_s x_1 \cdots x_i \in a q^{s-1} M + \mathfrak{m} q^s M,$$

therefore $r_i \in \mathfrak{m}$.

If $n \le s$ then it is easy to see that $y_{i,n} = a^i x_{i+1} \cdots x_n m$, $i = 0, \ldots, n$ is part of a minimal system of generators of $q^n M$. In fact multiplying by $x_{n+1} \cdots x_s$ we map the elements $\{y_{0,n}, \ldots, y_{n,n}\}$ of $q^n M$ onto $\{y_0, \ldots, y_n\}$ which is part of a minimal set of generators of $q^s M$ and (1) is proved. We remark now that (2) is a trivial consequence of the definition of $s(\mathbb{M})$ and of the fact that a is M-regular.   □

Here we extend Herzog–Waldi and Sally's results to our general setting.

Proposition 2.7. *Let $\mathbb{M}$ be the q-adic filtration of a Cohen–Macaulay module M of dimension one. Let p be an integer such that $qM \subseteq \mathfrak{m}^p M$. Then*

(1) $H_{\mathbb{M}}(n) \ge n + p$ for every $n \le s(\mathbb{M})$

(2) $s(\mathbb{M}) \le e_0(\mathbb{M}) - p$.

Proof. Since

$$q^n M \supseteq \mathfrak{m} q^n M \supseteq \mathfrak{m}^2 q^n M \supseteq \cdots \supseteq \mathfrak{m}^p q^n M \supseteq q^{n+1} M$$

we get

$$\begin{aligned} H_{\mathbb{M}}(n) = \lambda(q^n M / q^{n+1} M) &= \mu(q^n M) + \mu(\mathfrak{m} q^n M) + \cdots + \mu(\mathfrak{m}^{p-1} q^n M) \\ &+ \lambda(\mathfrak{m}^p q^n M / q^{n+1} M). \end{aligned}$$

Now, by Theorem 2.2, we get $H_{\mathbb{M}}(n) \ge n + p$ if $n \le s(\mathbb{M})$.

Since $H_{\mathbb{M}}(n) \le e_0(\mathbb{M})$, as consequence of (1) we get $H_{\mathbb{M}}(e_0(\mathbb{M}) - p) = e_0(\mathbb{M})$. Hence

$$s(\mathbb{M}) \le e_0(\mathbb{M}) - p.$$

□

From the proof of the above result we remark that if $\mathfrak{m}^p q^n M \ne q^{n+1} M$, then:

(a) $H_{\mathbb{M}}(n) > n + p$ if $n \le s(\mathbb{M})$

(b) $s(\mathbb{M}) \le e_0(\mathbb{M}) - p - 1$.

As a consequence of Propositions 2.7 and 2.3, we obtain the following proposition which refines a classical result proved by Kirby for the maximal ideal in [60] and extended to the $\mathfrak{m}$-primary ideals in [83].

Proposition 2.8. *Let $\mathbb{M}$ be the q-adic filtration of a module M of dimension one. Let p be an integer such that $qM \subseteq \mathfrak{m}^p M$. Then*

$$e_1(\mathbb{M}) - e_1(\mathbb{N}) \le \binom{e_0(\mathbb{M}) - p + 1}{2}.$$

If $e_0(\mathbb{M}) \ne e_0(\mathfrak{m}M)$, then

$$e_1(\mathbb{M}) - e_1(\mathbb{N}) \le \binom{e_0(\mathbb{M}) - p}{2}.$$

Proof. We recall that $e_0(\mathbb{M}) = e_0(\mathbb{M}^{sat})$ and, if $\mathfrak{q}M \subseteq \mathfrak{m}^p M$, clearly $\mathfrak{q}M^{sat} \subseteq \mathfrak{m}^p M^{sat}$. Then, by Proposition 2.3, we may assume M is a Cohen–Macaulay module and $s(\mathbb{M}) \geq 1$.

Since $e_1(\mathbb{M}) = \sum_{j \geq 0}(e_0(\mathbb{M}) - H_{\mathbb{M}}(j))$, by Proposition 2.7, it follows that

$$e_1(\mathbb{M}) = \sum_{j=0}^{e_0(\mathbb{M})-p}(e_0(\mathbb{M}) - H_{\mathbb{M}}(j)) \leq \sum_{j=0}^{e_0(\mathbb{M})-p}(e_0(\mathbb{M}) - j - p) = \binom{e_0(\mathbb{M}) - p + 1}{2}.$$

The last assertion is a consequence of the remark (a) following Proposition 2.7. □

2.3 The Higher Dimensional Case

We come now to the *higher dimensional* case. Here the strategy is to lower dimension by using superficial elements. We do not assume that M is Cohen–Macaulay, so we will get formulas containing a correction term which vanishes in the Cohen–Macaulay case.

If $J := (a_1, \ldots, a_r)$ is an $\mathbb{M}$-superficial sequence for $\mathfrak{q}$, let

$$\mathbb{N} := \{J^j M\}$$

be the J-adic filtration on M. It is not difficult to prove that $e_0(\mathbb{M}) = e_0(\mathbb{N})$.

If M is Cohen–Macaulay, then $e_i(\mathbb{N}) = 0$ for every $i \geq 1$, so that these integers are good candidates for being correction terms when the Cohen–Macaulay assumption does not hold. Concerning $e_1(\mathbb{M})$, we can say that $e_1(\mathbb{N})$ is the penalty for the lack of that condition. This character was already used by Goto in studying the Buchsbaum case. If M is a generalized Cohen–Macaulay module, then following [31], we have

$$e_1(\mathbb{N}) \geq -\sum_{i=1}^{r-1}\binom{r-2}{i-1}\lambda(H_{\mathfrak{m}}^i(M))$$

with equality if M is Buchsbaum. Hence if M is Buchsbaum, then $e_1(\mathbb{N})$ is independent of J. Very recently, under suitable assumptions and when $M = A$, Goto and Ozeki proved that if $e_1(\mathbb{N})$ is independent of J, then A is Buchsbaum (see [35]).

The following Lemma is the key to our investigation. It is due to David Conti.

Lemma 2.4. *Let $M_1, \ldots, M_d$ be A-modules of dimension r, let $\mathfrak{q}$ be an $\mathfrak{m}$-primary ideal of A and $\mathbb{M}_1 = \{M_{1,j}\}, \ldots, \mathbb{M}_d = \{M_{d,j}\}$ be good $\mathfrak{q}$-filtrations of $M_1, \ldots, M_d$ respectively. Then we can find a sequence of elements $a_1, \ldots, a_r$ which are $\mathbb{M}_i$-superficial for $\mathfrak{q}$ for every $i = 1, \ldots, d$.*

If $r \geq 2$ and $M_1 = \cdots = M_d = M$, then for every $1 \leq i \leq j \leq d$ we have

$$e_1(\mathbb{M}_i) - e_1(\mathbb{M}_i/a_1 M) = e_1(\mathbb{M}_j) - e_1(\mathbb{M}_j/a_1 M).$$

Proof. We have a filtration of the module $\oplus_{i=1}^{d} \mathbb{M}_i$

$$\oplus_{i=1}^{d} M_i \supseteq \oplus_{i=1}^{d} M_{i,1} \supseteq \oplus_{i=1}^{d} M_{i,2} \supseteq \cdots \supseteq \oplus_{i=1}^{d} M_{i,j} \supseteq \cdots$$

which we denote by $\oplus_{i=1}^{d} \mathbb{M}_i$. It is clear that this is a good q-filtration on $\oplus_{i=1}^{d} M_i$. Let us choose a $\oplus_{i=1}^{d} \mathbb{M}_i$-superficial sequence $\{a_1, \ldots, a_r\}$ for q. Then it is easy to see that $\{a_1, \ldots, a_r\}$ is a sequence of M_i-superficial elements for I for every $i = 1, \ldots, d$. This proves the first assertion. As for the second one, we know that

$$e_1(\mathbb{M}_i) - e_1(\mathbb{M}_i/a_1 M) = \begin{cases} \lambda(0 :_M a_1) & \text{if } r = 2 \\ 0 & \text{if } r \geq 3 \end{cases}$$

from which the conclusion follows.   □

We start with the extension of Proposition 2.4 to the higher dimensional case. To this end, given a good q-filtration $\mathbb{M} = \{M_j\}_{j \geq 0}$ of M and an ideal J generated by a maximal sequence of $\mathbb{M}$-superficial elements for q, we let for every $j \geq 0$,

$$v_j(\mathbb{M}) := \lambda(M_{j+1}/JM_j). \tag{2.7}$$

When M is one-dimensional and Cohen–Macaulay one has

$$v_j(\mathbb{M}) = u_j(\mathbb{M})$$

where the u_j's are defined as in (2.3). In the classical setting the following result is due to Huckaba and Marley.

Proposition 2.9. *Let $\mathbb{M}$ be a good q-filtration of a module M of dimension r and J an ideal generated by a maximal sequence of $\mathbb{M}$-superficial elements for q; then we have*

$$e_1(\mathbb{M}) - e_1(\mathbb{N}) \leq \sum_{j \geq 0} v_j(\mathbb{M}).$$

Proof. If $\dim(M) = 1$, then we can apply Proposition 2.4. Let $\dim(M) \geq 2$; by Lemma 2.4, we can find a minimal system of generators $\{a_1, \ldots, a_r\}$ of J such that a_1 is $\mathbb{N}$-superficial for J and $\{a_1, \ldots, a_r\}$ is a sequence of $\mathbb{M}$-superficial elements for q. The module $M/a_1 M$ has dimension $r - 1$ and $\mathbb{M}/a_1 M$ is a good q-filtration on it. Further it is clear that $a_2, \ldots, a_r$, is a maximal sequence of $\mathbb{M}/a_1 M$-superficial elements for q. Hence, if we let K be the ideal generated by $a_2, \ldots, a_r$, then by using induction on $\dim(M)$, we get

$$\begin{aligned} e_1(\mathbb{M}) - e_1(\mathbb{N}) &= e_1(\mathbb{M}/a_1 M) - e_1(\mathbb{N}/a_1 M) \\ &\leq \sum_{j \geq 0} \lambda\left((M_{j+1} + a_1 M)/(KM_j + a_1 M)\right) \\ &= \sum_{j \geq 0} \lambda\left((M_{j+1} + a_1 M)/(JM_j + a_1 M)\right) \end{aligned}$$

$$= \sum_{j \geq 0} \lambda \left(M_{j+1}/(JM_j + (a_1 M \cap M_{j+1})) \right)$$

$$\leq \sum_{j \geq 0} \lambda \left(M_{j+1}/JM_j \right)$$

$\square$

In the classical case, when $M = A$, $\mathbb{M} = \{\mathfrak{q}^j\}$ with $\mathfrak{q}$ an $\mathfrak{m}$-primary ideal of the r-dimensional Cohen–Macaulay local ring A, the above inequality is due to S. Huckaba (see [49]). Huckaba also proved that equality holds if and only if the associated graded ring has depth at least $r - 1$. We will extend this result in the next section (see Theorems 2.5 and 2.6).

We move now to the extension of Proposition 2.8. If $(A, \mathfrak{m})$ is a Cohen–Macaulay local ring, we recall that Kirby (see [60]) proved

$$e_1(\mathfrak{m}) \leq \binom{e_0(\mathfrak{m})}{2}$$

If $\mathfrak{q}$ is an $\mathfrak{m}$-primary ideal, the result has been extended in [83]. In particular if $e_0(\mathfrak{q}) \neq e_0(\mathfrak{m})$, then

$$e_1(\mathfrak{q}) \leq \binom{e_0(\mathfrak{q}) - 1}{2}$$

We improve the above results by using the machinery already introduced.

Proposition 2.10. *Let $\mathbb{M}$ be the $\mathfrak{q}$-adic filtration on a module M of dimension r. Let p be an integer such that $\mathfrak{q}M \subseteq \mathfrak{m}^p M$. Then*

$$e_1(\mathbb{M}) - e_1(\mathbb{N}) \leq \binom{e_0(\mathbb{M}) - p + 1}{2}.$$

Proof. We proceed by induction on $r = \dim M$. If $r = 1$ the result follows by Proposition 2.8. If $r \geq 2$, as before, we can find a minimal system of generators $\{a_1, \cdots, a_r\}$ of J such that a_1 is $\mathbb{N}$-superficial for J and $\{a_1, \cdots, a_r\}$ is a sequence of $\mathbb{M}$-superficial elements for $\mathfrak{q}$. The module $M/a_1 M$ has dimension $r - 1$ and $\mathbb{M}/a_1 M$ is a good $\mathfrak{q}$-filtration on it. Now $e_1(\mathbb{M}) - e_1(\mathbb{N}) = e_1(\mathbb{M}/a_1 M) - e_1(\mathbb{N}/a_1 M)$, $e_0(\mathbb{M}) = e_0(\mathbb{M}/a_1 M)$ and $\mathfrak{q}M/a_1 M \subseteq \mathfrak{m}^p M/a_1 M$. Hence the result follows by the inductive assumption. $\square$

For completeness we recall that, by using a deeper investigation, for local Cohen–Macaulay rings of embedding dimension b and dimension r, Elias in [20, Theorem 1.6] proved

$$e_1(\mathfrak{m}) \leq \binom{e_0(\mathfrak{m})}{2} - \binom{b - r}{2}.$$

An easier approach was presented by the authors in [84] where the result was proved for any $\mathfrak{m}$-primary ideal $\mathfrak{q}$.

Theorem 2.3. *Let* $(A, \mathfrak{m})$ *be a Cohen–Macaulay local ring of dimension r and let $\mathfrak{q}$ be an $\mathfrak{m}$-primary ideal in A. Then*

$$e_1(\mathfrak{q}) \leq \binom{e_0(\mathfrak{q})}{2} - \binom{\mu(\mathfrak{q}) - r}{2} - \lambda(A/\mathfrak{q}) + 1.$$

Notice that in the particular case of an $\mathfrak{m}$-primary ideal $\mathfrak{q} \subseteq \mathfrak{m}^2$ a nice proof was produced by Elias in [24].

We move now to the higher dimensional case of Proposition 2.5. This improves *Northcott's inequality* to filtrations of a module which is not necessarily Cohen–Macaulay.

Theorem 2.4. *Let* $\mathbb{M} = \{M_j\}_{j \geq 0}$ *be a good $\mathfrak{q}$-filtration of a module M of dimension r. If $s \geq 1$ is an integer and J is an ideal generated by a maximal $\mathbb{M}$-superficial sequence for $\mathfrak{q}$, then we have:*

$$e_1(\mathbb{M}) - e_1(\mathbb{N}) \geq s e_0(\mathbb{M}) - \lambda(M/M_{s-1}) - \lambda(M/M_s + JM).$$

Proof. If $r = 1$, by Proposition 2.5 we get

$$e_1(\mathbb{M}) - e_1(\mathbb{N}) \geq s e_0(\mathbb{M}) - \lambda(M/M_s).$$

Hence we must prove that

$$s e_0(\mathbb{M}) - \lambda(M/M_s) \geq s e_0(\mathbb{M}) - \lambda(M/M_{s-1}) - \lambda(M/M_s + JM).$$

This is equivalent to proving

$$\lambda(M/M_{s-1}) \geq \lambda(M_s + JM/M_s) = \lambda(JM/JM \cap M_s).$$

Since $J = (a)$ is a principal ideal and $aM_{s-1} \subseteq aM \cap M_s$, we have a surjection

$$M/M_{s-1} \xrightarrow{a} aM/aM \cap M_s$$

and the conclusion follows in this case.

Let $r \geq 2$; by using the above remark, we can find a minimal system of generators $\{a_1, \ldots, a_r\}$ of J such that a_1 is $\mathbb{N}$-superficial for J and $\{a_1, \ldots, a_r\}$ is a sequence of $\mathbb{M}$-superficial elements for $\mathfrak{q}$.

The module M/a_1M has dimension $r-1$ and $\mathbb{M}/a_1M$ is a good $\mathfrak{q}$-filtration on it. Furthermore it is clear that $a_2, \ldots, a_r$ is a maximal sequence of $\mathbb{M}/a_1M$-superficial elements for $\mathfrak{q}$. Hence, if we let K be the ideal generated by $a_2, \ldots, a_r$ and $\mathbb{K}$ the K-adic filtration on M/a_1M, then by induction, and after a little standard computation, we get

$$e_1(\mathbb{M}/a_1M) - e_1(\mathbb{K}) \geq s e_0(\mathbb{M}/a_1M) - \lambda(M/M_{s-1} + a_1M) - \lambda(M/M_s + JM).$$

Since $e_0(\mathbb{M}/a_1 M) = e_0(\mathbb{M})$, $\mathbb{N}/a_1 M = \mathbb{K}$ and

$$e_1(\mathbb{M}) - e_1(\mathbb{M}/aM) = e_1(\mathbb{N}) - e_1(\mathbb{N}/aM),$$

we finally get

$$e_1(\mathbb{M}) - e_1(\mathbb{N}) \geq s e_0(\mathbb{M}) - \lambda(M/M_{s-1} + a_1 M) - \lambda(M/M_s + JM)$$
$$\geq s e_0(\mathbb{M}) - \lambda(M/M_{s-1}) - \lambda(M/M_s + JM)$$

which gives the conclusion. $\square$

Remark 2.1. Let us apply our theorem to the very particular case when $M_j = \mathfrak{q}^j$ for every $j \geq 0$ and $\mathfrak{q}$ is a primary ideal of A. It is well known that if J is any minimal reduction of $\mathfrak{q}$, then we have an injection

$$J/J\mathfrak{m} \to \mathfrak{q}/\mathfrak{q}\mathfrak{m}$$

which proves that any minimal system of generators of J is part of a minimal system of generators of $\mathfrak{q}$. Furthermore J can be minimally generated by a maximal sequence of $\mathbb{M}$-superficial elements for $\mathfrak{q}$. Hence we can apply the above theorem to get:

- $s = 1$

$$e_1(\mathfrak{q}) - e_1(J) \geq e_0(\mathfrak{q}) - \lambda(A/\mathfrak{q})$$

which is exactly Theorem 3.1 in [31].

- $s = 2$

$$e_1(\mathfrak{q}) - e_1(J) \geq 2 e_0(\mathfrak{q}) - \lambda(A/\mathfrak{q}) - \lambda(A/\mathfrak{q}^2 + J).$$

This means

$$2 e_0(\mathfrak{q}) - e_1(\mathfrak{q}) + e_1(J) \leq 2\lambda(A/\mathfrak{q}) + \lambda(\mathfrak{q}/\mathfrak{q}^2 + J).$$

Since $r = \lambda(J/J\mathfrak{m})$, if we let $t := \lambda(\mathfrak{q}/\mathfrak{q}\mathfrak{m}) - r$, we can find elements $x_1, \cdots, x_t \in \mathfrak{q}$ such that $\mathfrak{q} = J + (x_1, \ldots, x_t)$. Hence the canonical map

$$\varphi : (A/\mathfrak{q})^t \to \mathfrak{q}/(\mathfrak{q}^2 + J)$$

given by $\varphi(\overline{a_1}, \ldots, \overline{a_t}) = \sum \overline{a_i x_i}$ is surjective and we get

$$2 e_0(\mathfrak{q}) - e_1(\mathfrak{q}) + e_1(J) \leq (t + 2)\lambda(A/\mathfrak{q})$$

which is Proposition 3.7 in [12].

When $M = A$ is Cohen–Macaulay and $\mathbb{M}$ is a Hilbert filtration on A, which means that $M_j = I_j$ with I_j ideals in A, $I_0 = A$, I_1 is $\mathfrak{m}$-primary and $\mathbb{M}$ is I_1-good, Guerrieri and Rossi proved in [38] the following formula:

$$e_1(\mathbb{M}) \geq 2e_0(\mathbb{M}) - \left(\lambda(I_1/I_2) - r\lambda(A/I_1) + \lambda((I_2 \cap J)/JI_1) + 2\lambda(A/I_1)\right).$$

If we apply the above theorem in this situation, we get

$$e_1(\mathbb{M}) - e_1(\mathbb{N}) \geq 2e_0(\mathbb{M}) - \lambda(A/I_1) - \lambda(A/I_2 + J).$$

Since A is Cohen–Macaulay, every superficial sequence is a regular sequence in A and thus $e_1(\mathbb{N}) = 0$ and $r\lambda(A/I_1) = \lambda(J/JI_1)$. Then, by easy computation, we can see that the two bounds coincide.

We now want to extend Proposition 2.6 to the higher dimensional case. We recall that given a good $\mathfrak{q}$-filtration $\mathbb{M}$ of the r-dimensional module M, we can consider the ideal J generated by a maximal $\mathbb{M}$-superficial sequence for $\mathfrak{q}$, and we are interested in the study of two related filtrations on M: the J-adic filtration $\mathbb{N} := \{J^j M\}$ already defined and the filtration $\mathbb{E}$ given by

$$\mathbb{E}: \quad M \supseteq M_1 \supseteq JM_1 \supseteq J^2 M_1 \supseteq \cdots \supseteq J^j M_1 \supseteq J^{j+1} M_1 \supseteq \cdots$$

In the following for any good $\mathfrak{q}$-filtration $\mathbb{M}$ and for every ideal J generated by a maximal $\mathbb{M}$-superficial sequence for $\mathfrak{q}$, we may associate the two good J-filtrations $\mathbb{N}$ and $\mathbb{E}$.

As in the remark before Theorem 2.4, by using Proposition 1.2, we can easily see that

$$e_1(\mathbb{E}) - e_1(\mathbb{E}/aM) = e_1(\mathbb{N}) - e_1(\mathbb{N}/aM).$$

Notice that in general $e_1(\mathbb{N})$ and $e_1(\mathbb{E})$ depend on J (see [28, 29, 35, 111]).

Proposition 2.11. *Let $\mathbb{M}$ be a good $\mathfrak{q}$-filtration of a module M of dimension r and let J be an ideal generated by a maximal $\mathbb{M}$-superficial sequence for $\mathfrak{q}$. Then we have*

$$e_1(\mathbb{E}) - e_1(\mathbb{N}) \geq e_0(\mathbb{M}) - h_0(\mathbb{M}) + \lambda(M_1 + H^0(M)/M_1).$$

Proof. If $r = 1$ we apply Proposition 2.6. Let $r \geq 2$; as before we can find an element $a \in J$ which is superficial for $\mathbb{N}$ and $\mathbb{E}$. Then we have

$$e_1(\mathbb{E}) - e_1(\mathbb{N}) = e_1(\mathbb{E}/aM) - e_1(\mathbb{N}/aM))$$
$$\geq e_0(\mathbb{M}/aM) - h_0(\mathbb{M}/aM) + \lambda\left((M_1 + aM/aM) + H^0(M/aM))/(M_1 + aM/aM)\right)$$
$$= e_0(\mathbb{M}) - h_0(\mathbb{M}) + \lambda\left(aM :_M \mathfrak{m}^n + M_1/M_1\right) \geq e_0(\mathbb{M}) - h_0(\mathbb{M}) + \lambda(M_1 + H^0(M)/M_1).$$

$\square$

2.4 The Border Cases

The aim of this section is the study of the extremal cases with respect to the inequalities proved in the above section. With few exceptions we assume M is a filtered Cohen–Macaulay module.

The following result, first proved in the classical case by S. Huckaba in [49], has been reconsidered and extended in a series of recent papers (see [17, 41, 53, 113]), where, unfortunately, the original heavy homological background was still essential. We remark that the statements involving the Hilbert coefficients e_j with $j \geq 2$ are new even in the classical case, except for the bound on e_2 which had been already proved in [14].

We recall that, given a good $\mathfrak{q}$-filtration $\mathbb{M}$ of the module M and an ideal J generated by a maximal sequence of $\mathbb{M}$-superficial elements for $\mathfrak{q}$, we denote by $v_j(\mathbb{M})$ the non negative integers

$$v_j(\mathbb{M}) := \lambda(M_{j+1}/JM_j).$$

In Lemma 2.2 we proved that, if M is one-dimensional and Cohen–Macaulay, then for every $i \geq 0$

$$e_i(\mathbb{M}) = \sum_{j \geq i-1} \binom{j}{i-1} v_j(\mathbb{M}),$$

while, in Proposition 2.9, we proved that if M is Cohen–Macaulay then

$$e_1(\mathbb{M}) \leq \sum_{j \geq 0} v_j(\mathbb{M}).$$

Theorem 2.5. *Let $\mathbb{M}$ be a good $\mathfrak{q}$-filtration of a Cohen–Macaulay module M of dimension r and let J be an ideal generated by a maximal $\mathbb{M}$-superficial sequence for $\mathfrak{q}$. Then we have:*

(a) $e_1(\mathbb{M}) \leq \sum_{j \geq 0} v_j(\mathbb{M})$

(b) $e_2(\mathbb{M}) \leq \sum_{j \geq 0} j v_j(\mathbb{M})$.

(c) The following conditions are equivalent:

    (1) depth $gr_{\mathbb{M}}(M) \geq r - 1$.

    (2) $e_i(\mathbb{M}) = \sum_{j \geq i-1} \binom{j}{i-1} v_j(\mathbb{M})$ *for every* $i \geq 1$.

    (3) $e_1(\mathbb{M}) = \sum_{j \geq 0} v_j(\mathbb{M})$.

    (4) $e_2(\mathbb{M}) = \sum_{j \geq 0} j v_j(\mathbb{M})$.

    (5) $P_{\mathbb{M}}(z) = \frac{\lambda(M/M_1) + \sum_{j \geq 0} (v_j(\mathbb{M}) - v_{j+1}(\mathbb{M})) z^{j+1}}{(1-z)^r}$.

Proof. Let $J = (a_1,\ldots,a_r)$ and $\mathfrak{a} = (a_1,\ldots,a_{r-1})$; we first remark that, by Lemma 1.3 and Theorem 1.1, depth $gr_{\mathbb{M}}(M) \geq r - 1$ if and only if $M_{j+1} \cap \mathfrak{a}M = \mathfrak{a}M_j$ for every $j \geq 0$. Further we have

$$v_j(\mathbb{M}) = v_j(\mathbb{M}/\mathfrak{a}M) + \lambda\,(M_{j+1} \cap \mathfrak{a}M + JM_j/JM_j),$$

hence $v_j(\mathbb{M}) \geq v_j(\mathbb{M}/\mathfrak{a}M)$ and equality holds if and only if $M_{j+1} \cap \mathfrak{a}M \subseteq JM_j$. This is certainly the case when $M_{j+1} \cap \mathfrak{a}M = \mathfrak{a}M_j$.

Hence, if depth $gr_{\mathbb{M}}(M) \geq r - 1$, then $v_j(\mathbb{M}) = v_j(\mathbb{M}/\mathfrak{a}M)$ for every $j \geq 0$. By induction on j, we can prove that the converse holds. Namely $M_1 \cap \mathfrak{a}M = \mathfrak{a}M$ and, if $j \geq 1$, then we have

$$\begin{aligned}
M_{j+1} \cap \mathfrak{a}M \subseteq JM_j \cap \mathfrak{a}M &= (\mathfrak{a}M_j + a_r M_j) \cap \mathfrak{a}M \\
&= \mathfrak{a}M_j + (a_r M_j \cap \mathfrak{a}M) \subseteq \mathfrak{a}M_j + a_r(M_j \cap \mathfrak{a}M) \\
&= \mathfrak{a}M_j + a_r \mathfrak{a}M_{j-1} = \mathfrak{a}M_j.
\end{aligned}$$

where $a_r M_j \cap \mathfrak{a}M \subseteq a_r(M_j \cap \mathfrak{a}M)$ because a_r is regular modulo $\mathfrak{a}M$, while $M_j \cap \mathfrak{a}M = \mathfrak{a}M_{j-1}$ follows by induction.

Since $M/\mathfrak{a}M$ is Cohen–Macaulay of dimension one, we get

$$e_1(\mathbb{M}) = e_1(\mathbb{M}/\mathfrak{a}M) = \sum_{j \geq 0} v_j(\mathbb{M}/\mathfrak{a}M) \leq \sum_{j \geq 0} v_j(\mathbb{M}).$$

Equality holds if and only if depth $gr_{\mathbb{M}}(M) \geq r - 1$. This proves (a) once more and moreover gives the equivalence between (1) and (3) in (c). By using (2.4) and Proposition 1.2 this also gives the equivalence between (1) and (5) in (c).

Now, if $\mathfrak{b}$ is the ideal generated by $a_1,\ldots,a_{r-2}$, then, as before, we get

$$e_2(\mathbb{M}) = e_2(\mathbb{M}/\mathfrak{b}M) \leq e_2(\mathbb{M}/\mathfrak{a}M) = \sum_{j \geq 1} j v_j(\mathbb{M}/\mathfrak{a}M) \leq \sum_{j \geq 1} j v_j(\mathbb{M}).$$

This proves (b) and $4 \Longrightarrow 1$. To complete the proof of the theorem, we need only to show that $1 \Longrightarrow 2$. If depth $gr_{\mathbb{M}}(M) \geq r - 1$, then $\mathbb{M}$ and $\mathbb{M}/\mathfrak{a}M$ have the same h-polynomial; this implies that for every $i \geq 1$ we have

$$e_i(\mathbb{M}) = e_i(\mathbb{M}/\mathfrak{a}M) = \sum_{j \geq i-1} \binom{j}{i-1} v_j(\mathbb{M}/\mathfrak{a}M) = \sum_{j \geq i-1} \binom{j}{i-1} v_j(\mathbb{M}).$$

$\square$

In the above result the equality in (b) does not force $gr_{\mathbb{M}}(M)$ to be Cohen–Macaulay. In fact we will see later that in a two-dimensional local Cohen–Macaulay ring, e_2 can be zero, but depth $gr_{\mathbb{M}}(M) = 0$ (see Example 3.3).

We recall that Proposition 2.9 extends Huckaba's inequality without assuming the Cohen–Macaulayness of M. Namely we proved that

$$e_1(\mathbb{M}) - e_1(\mathbb{N}) \leq \sum_{j \geq 0} v_j(\mathbb{M})$$

where $\mathbb{N}$ is the J-adic filtration on M.

If we do not assume that M is Cohen–Macaulay, we are able to handle the equality only for the m-adic filtration on A. Surprisingly in [85, Theorem 2.13], the authors proved that the equality in Proposition 2.9 forces the ring A itself to be Cohen–Macaulay and hence, by Theorem 2.5, $gr_{\mathfrak{m}}(A)$ to have almost maximal depth. The result is the following.

Theorem 2.6. *Let $(A, \mathfrak{m})$ be a local ring of dimension $r \geq 1$ and let J be the ideal generated by a maximal $\mathfrak{m}$-superficial sequence. The following conditions are equivalent:*

(1) $e_1(\mathfrak{m}) - e_1(J) = \sum_{j \geq 0} v_j(\mathfrak{m})$.

(2) A is Cohen–Macaulay and depth $gr_{\mathfrak{m}}(A) \geq r - 1$.

Proof. If A is Cohen–Macaulay, then $e_1(J) = 0$ and, by the above result, we find that 2) implies 1).

We prove now that 1) implies 2) by induction on r. If $r = 1$, the result follows by Proposition 2.4 since $W \subseteq \mathfrak{m}$. Let $r \geq 2$; by Lemma 2.4 we can find a minimal basis $\{a_1, \ldots, a_r\}$ of J such that a_1 is J-superficial, $\{a_1, \ldots, a_r\}$ is an m-superficial sequence and $e_1(\mathfrak{m}) - e_1(J) = e_1(\mathfrak{m}/(a_1)) - e_1(J/(a_1))$. Now $A/(a_1)$ is a local ring of dimension $d - 1$ and $J/(a_1)$ is generated by a maximal $\mathfrak{m}/(a_1)$-superficial sequence. We can then apply Proposition 2.9 to get $\sum_{j \geq 0} v_j(\mathfrak{m}) = e_1(\mathfrak{m}) - e_1(J) = e_1(\mathfrak{m}/(a_1)) - e_1(J/(a_1)) \leq \sum_{j \geq 0} v_j(\mathfrak{m}/(a_1)) \leq \sum_{j \geq 0} v_j(\mathfrak{m})$.

This implies

$$e_1(\mathfrak{m}/(a_1)) - e_1(J/(a_1)) = \sum_{j \geq 0} v_j(\mathfrak{m}/(a_1))$$

which, by the inductive assumption, implies that $A/(a_1)$ is Cohen–Macaulay. By Lemma 1.5 A is Cohen–Macaulay so that $e_1(J) = 0$ and then $e_1(\mathfrak{m}) = \sum_{j \geq 0} v_j(\mathfrak{m})$; this implies depth $gr_{\mathfrak{m}}(A) \geq r - 1$ and the result is proved. $\square$

Remark 2.2. As the reader can see, the above result was presented for the m-adic filtration. Actually a more general statement holds. In order to apply Proposition 2.4 it is enough to require a filtration $\mathbb{M}$ such that $W \subseteq M_1$.

Since $e_1(J) \leq 0$ (see [27–29]), as a trivial consequence of the above result we have the following interesting extension of Huckaba-Marley's result where the Cohen–Macaulayness of A is assumed.

Corollary 2.4. *Let* $(A, \mathfrak{m})$ *be a local ring of dimension* $r \geq 1$ *and let* J *be the ideal generated by a maximal* $\mathfrak{m}$-*superficial sequence. Then*

$$e_1(\mathfrak{m}) \leq \sum_{j \geq 0} v_j(\mathfrak{m}).$$

Moreover, the following conditions are equivalent:

(1) $e_1(\mathfrak{m}) = \sum_{j \geq 0} v_j(\mathfrak{m})$.

(2) A *is Cohen–Macaulay and* $\operatorname{depth} gr_{\mathfrak{m}}(A) \geq r - 1$.

In the Cohen–Macaulay case we describe now another set of numerical characters of the filtered module M, which are important in the study of the Hilbert coefficients.

Let $\mathbb{M} = \{M_j\}_{j \geq 0}$ be a good q-filtration of M and J an ideal generated by an $\mathbb{M}$-superficial sequence for q; then, for every $j \geq 0$, we let

$$w_j(\mathbb{M}) := \lambda(M_{j+1} + JM/JM) = \lambda(M_{j+1}/M_{j+1} \cap JM). \tag{2.8}$$

Since

$$JM_j \subseteq JM \cap M_{j+1} \subseteq M_{j+1},$$

we get

$$v_j(\mathbb{M}) = w_j(\mathbb{M}) + \lambda(JM \cap M_{j+1}/JM_j) \tag{2.9}$$

The length of the abelian group $JM \cap M_{j+1}/JM_j$ will be denoted by $vv_j(\mathbb{M})$ since these groups are the homogeneous components of the Valabrega–Valla module

$$VV(\mathbb{M}) := \bigoplus_{j \geq 0} (JM \cap M_{j+1}/JM_j)$$

of $\mathbb{M}$ with respect to J, as defined in [109, Chap. 5]. For example one has

$$vv_0(\mathbb{M}) = 0, \quad vv_1(\mathbb{M}) = \lambda(JM \cap M_2/JM_1)$$

and since $M_{j+1} = JM_j$ for large j, $vv_j(\mathbb{M}) = 0$ for $j \gg 0$. It follows that, in the case of the m-adic filtration $\{M_j = \mathfrak{m}^j\}$ of the Cohen–Macaulay ring A, one has $vv_1(\mathbb{M}) = 0$ by the analytic independence of a maximal regular sequence.

The significance of $VV(\mathbb{M})$ lies in the fact that, if M is Cohen–Macaulay, by Valabrega–Valla, $gr_{\mathbb{M}}(M)$ is Cohen–Macaulay if and only if $VV(\mathbb{M}) = 0$.

As a consequence we have that $gr_{\mathbb{M}}(M)$ is Cohen–Macaulay if and only if $v_j(\mathbb{M}) = w_j(\mathbb{M})$ for every $j \geq 0$.

In the one-dimensional Cohen–Macaulay case, we have

$$e_1(\mathbb{M}) = \sum_{j \geq 0} v_j(\mathbb{M}) = \sum_{j \geq 0} w_j(\mathbb{M}) + \sum_{j \geq 0} vv_j(\mathbb{M}), \tag{2.10}$$

so that $\sum_{j \geq 0} w_j(\mathbb{M}) \leq e_1(\mathbb{M})$ and equality holds if and only if $gr_{\mathbb{M}}(M)$ is Cohen–Macaulay.

This result can be extended to higher dimensions; we first need to remark that the integers w_j do not change upon reduction modulo a superficial sequence. Namely if $a \in J$, then

$$w_j(\mathbb{M}/aM) = \lambda\left((M_{j+1} + aM/aM) + (JM/aM)/(JM/aM)\right) =$$

$$= \lambda\left((M_{j+1} + JM/aM)/(J(M/aM))\right) = \lambda(M_{j+1} + JM/JM) = w_j(\mathbb{M}).$$

Theorem 2.7. *Let $\mathbb{M}$ be a good $\mathfrak{q}$-filtration of the Cohen–Macaulay module M of dimension $r \geq 1$ and let J be an ideal generated by a maximal $\mathbb{M}$-superficial sequence for $\mathfrak{q}$. Then we have:*

(a) $e_1(\mathbb{M}) \geq \sum_{j \geq 0} w_j(\mathbb{M})$

(b) $e_1(\mathbb{M}) = \sum_{j \geq 0} w_j(\mathbb{M})$ *if and only if* $gr_{\mathbb{M}}(M)$ *is Cohen–Macaulay.*

Proof. If $r = 1$ the result follows from (2.10). We assume $r \geq 2$ and let $a_1, \ldots, a_r$ be a superficial sequence which generates J. Denote $\mathfrak{a} = (a_1, \ldots, a_{r-1})$, then we have

$$e_1(\mathbb{M}) = e_1(\mathbb{M}/\mathfrak{a}M) = \sum_{j \geq 0} v_j(\mathbb{M}/\mathfrak{a}M) = \sum_{j \geq 0} w_j(\mathbb{M}/\mathfrak{a}M) + \sum_{j \geq 0} vv_j(\mathbb{M}/\mathfrak{a}M) =$$

$$= \sum_{j \geq 0} w_j(\mathbb{M}) + \sum_{j \geq 0} vv_j(\mathbb{M}/\mathfrak{a}M).$$

This proves (a) and also implies that $e_1(\mathbb{M}) = \sum_{j \geq 0} w_j(\mathbb{M})$ if and only if $VV(\mathbb{M}/\mathfrak{a}M) = 0$, if and only if $gr_{\mathbb{M}/\mathfrak{a}M}(M/\mathfrak{a}M)$ is Cohen–Macaulay. Hence (b) follows by Sally's machine. $\square$

As a corollary of Theorem 2.7, we get the following achievement which extends to our general setting the main result of A. Guerrieri in [36]. Here the proof is quite simple and is due to Cortadellas (see [17]).

Corollary 2.5. *Let $\mathbb{M}$ be a good $\mathfrak{q}$-filtration of the Cohen–Macaulay module M of dimension $r \geq 1$ and let J be an ideal generated by a maximal $\mathbb{M}$-superficial sequence for $\mathfrak{q}$. If $\lambda(VV(\mathbb{M})) = 1$, then depth $gr_{\mathbb{M}}(M) = r - 1$.*

Proof. Since $\lambda(VV(\mathbb{M})) = 1$, we have

$$\sum_{j \geq 0} w_j(\mathbb{M}) = \sum_{j \geq 0} v_j(\mathbb{M}) - 1.$$

Since for some $n \geq 0$ we have $M_{n+1} \cap JM \neq JM_n$, we have depth $gr_{\mathbb{M}}(M) \leq r - 1$, hence $e_1(\mathbb{M}) > \sum_{j \geq 0} w_j(\mathbb{M})$. We get

$$\sum_{j \geq 0} v_j(\mathbb{M}) \geq e_1(\mathbb{M}) > \sum_{j \geq 0} w_j(\mathbb{M}) = \sum_{j \geq 0} v_j(\mathbb{M}) - 1.$$

This implies $\sum_{j \geq 0} v_j(\mathbb{M}) = e_1(\mathbb{M})$ and the conclusion follows. $\square$

In the classical case, it was proved by Guerrieri in [37] that if $\lambda(q^2 \cap J/qJ) = = 2 = \lambda(VV(\mathbb{M}))$, then depth $gr_{\mathbb{M}}(M) \geq r - 2$, a result which was extended by Wang in [116], where he proved that if $\lambda(VV(\mathbb{M})) = 2$ then depth $gr_{\mathbb{M}}(M) \geq r - 2$. If $\lambda(q^2 \cap J/qJ) = \lambda(VV(\mathbb{M})) = 3$, Guerrieri and Rossi proved that depth $gr_{\mathbb{M}}(M) \geq r - 3$, provided A is Gorenstein. A conjecture of C. Huneke predicts that if $vv_j \leq 1$ for every j, then depth $gr_{\mathbb{M}}(M) \geq r - 1$. This is not true as shown by Wang. However Colomé and Elias proved that the condition $vv_j \leq 1$ for every j, implies depth $gr_{\mathbb{M}}(M) \geq r - 2$. Concerning this topic see also [8, 36, 39, 67, 116].

We want now to study the extremal case in Northcott's inequality. First we need to recall a lower bound for e_1 which was proved by Elias and Valla in [26]. Given a Cohen–Macaulay local ring $(A, \mathfrak{m})$, one has

$$e_1(\mathfrak{m}) \geq 2e_0(\mathfrak{m}) - h - 2, \tag{2.11}$$

where $h = \mu(\mathfrak{m}) - \dim(A)$ is the embedding codimension of A. Equality holds above if and only if the h-polynomial is short enough.

In our general setting we have the inequality given by Theorem 2.4, namely

$$e_1(\mathbb{M}) - e_1(\mathbb{N}) \geq s e_0(\mathbb{M}) - \lambda(M/M_{s-1}) - \lambda(M/M_s + JM).$$

When M is Cohen–Macaulay, we have $e_1(\mathbb{N}) = 0$ so that, if $s = 2$, we get

$$e_1(\mathbb{M}) \geq 2 e_0(\mathbb{M}) - \lambda(M/M_1) - \lambda(M/M_2 + JM).$$

If we let

$$h(\mathbb{M}) := \lambda(M_1/JM + M_2), \tag{2.12}$$

then we have

$$\begin{aligned} e_1(\mathbb{M}) &\geq 2 e_0(\mathbb{M}) - \lambda(M/M_1) - \lambda(M/M_2 + JM) \\ &= 2 e_0(\mathbb{M}) - h(\mathbb{M}) - 2 h_0(\mathbb{M}) \end{aligned} \tag{2.13}$$

a formula which extends (2.11) because $\lambda(\mathfrak{m}/J + \mathfrak{m}^2) = \mu(\mathfrak{m}) - \dim(A) = h$.

We remark that the integer $h(\mathbb{M})$ coincides with the embedding codimension in the case of the $\mathfrak{m}$-adic filtration. Further we have

$$h(\mathbb{M}) = h(\mathbb{M}/JM) = h_1(\mathbb{M}/JM) \tag{2.14}$$

and also

$$h(\mathbb{M}) = h_1(\mathbb{M}) + \lambda(M_2 \cap JM/JM_1). \tag{2.15}$$

The proof of the following theorem is exactly the same as the original given in [26] and [38]. We reproduce it here because is a typical example of the strategy to reduce dimension by using superficial elements and the Sally machine.

We recall that, when M is Cohen–Macaulay and $\dim(M) = 1$, we introduced the reduction number of the filtration $\mathbb{M}$ as the integer $s(\mathbb{M}) = \min\{j : H_{\mathbb{M}}(j) = e_0(\mathbb{M})\}$ and it turns out that it is also the degree of the h-polynomial of M, see (2.6).

In the following $s(\mathbb{M})$ will denote the *degree of the h-polynomial of M* (see Sect. 1.3 for the definition).

Theorem 2.8. *Let* $\mathbb{M}$ *be a good* $\mathfrak{q}$-*filtration of a Cohen–Macaulay module M and let J be an ideal generated by a maximal* $\mathbb{M}$-*superficial sequence for* $\mathfrak{q}$. *The following conditions are equivalent:*

(a) $e_1(\mathbb{M}) = 2e_0(\mathbb{M}) - 2h_0(\mathbb{M}) - h(\mathbb{M})$

(b) $s(\mathbb{M}) \leq 2$ *and* $M_2 \cap JM = JM_1$.

If either of the above conditions holds, then $gr_{\mathbb{M}}(M)$ *is Cohen–Macaulay.*

Proof. We prove that (b) implies (a). Since $M_2 \cap JM = JM_1$ and M is Cohen–Macaulay, we get $h(\mathbb{M}) = h_1(\mathbb{M})$. Since $s(\mathbb{M}) \leq 2$, we have

$$e_0(\mathbb{M}) = h_0(\mathbb{M}) + h_1(\mathbb{M}) + h_2(\mathbb{M}) \qquad e_1(\mathbb{M}) = h_1(\mathbb{M}) + 2h_2(\mathbb{M}).$$

Hence

$$\begin{aligned} 2e_0(\mathbb{M}) - 2h_0(\mathbb{M}) - h(\mathbb{M}) &= 2(h_0(\mathbb{M}) + h_1(\mathbb{M}) + h_2(\mathbb{M})) - 2h_0(\mathbb{M}) - h_1(\mathbb{M}) \\ &= h_1(\mathbb{M}) + 2h_2(\mathbb{M}) = e_1(\mathbb{M}). \end{aligned}$$

Let us prove the converse by induction on $r := \dim(M)$. If $r = 0$, then we have $P_{\mathbb{M}}(z) = \sum_{i=0}^{s} h_i(\mathbb{M})$ with $h_i(\mathbb{M}) \geq 0$ and where we let $s := s(\mathbb{M})$. Since $h(\mathbb{M}) = h_1(\mathbb{M})$, it is clear that $e_1(\mathbb{M}) \geq 2e_0(\mathbb{M}) - 2h_0(\mathbb{M}) - h(\mathbb{M})$ and if we have equality, then

$$h_3(\mathbb{M}) + 2h_4(\mathbb{M}) + \cdots + (s-2)h_s(\mathbb{M}) = 0,$$

which implies $s \leq 2$.

If $r \geq 1$, let $J = (a_1, \ldots, a_r)$ and $K := (a_1, \ldots, a_{r-1})$. Then we have

$$\begin{aligned} 2e_0(\mathbb{M}) - 2h_0(\mathbb{M}) - h(\mathbb{M}) = e_1(\mathbb{M}) &= e_1(\mathbb{M}/KM) \geq e_1(\mathbb{M}/JM) \\ &\geq 2e_0(\mathbb{M}/JM) - 2h_0(\mathbb{M}/JM) - h(\mathbb{M}/JM) \\ &= 2e_0(\mathbb{M}) - 2h_0(\mathbb{M}) - h(\mathbb{M}) \end{aligned}$$

where we used Proposition 1.2 several times. This gives

$$e_1(\mathbb{M}/KM) = e_1(\mathbb{M}/JM)$$

which, again by Proposition 1.2, implies depth $gr_{\mathbb{M}}(M/KM) = 1$. By Sally's machine, $gr_{\mathbb{M}}(M)$ is Cohen–Macaulay so that $s(\mathbb{M}) = s(\mathbb{M}/JM) \leq 2$. Finally, by Valabrega and Valla, $M_2 \cap JM = JM_1$, as wanted. $\qquad\square$

We collect in the following formula some of the results we proved when M is Cohen–Macaulay and J is an ideal generated by a maximal sequence of $\mathbb{M}$-superficial elements for $\mathfrak{q}$.

$$\begin{aligned}
e_1(\mathbb{M}) &\geq 2e_0(\mathbb{M}) - h(\mathbb{M}) - 2h_0(\mathbb{M}) \\
&= e_0(\mathbb{M}) - h_0(\mathbb{M}) + \lambda(JM + M_2/JM) \\
&= h_1(\mathbb{M}) + \lambda(M_2/JM_1) + \lambda(M_2/JM \cap M_2).
\end{aligned} \tag{2.16}$$

Here the first inequality comes from Theorem 2.4 with $s = 1$, the first equality comes from the identities

$$e_0(\mathbb{M}) = \lambda(M/JM), \quad h_0(\mathbb{M}) = \lambda(M/M_1), \quad h(\mathbb{M}) = \lambda(M_1/JM + M_2)$$

and, finally, the last equality is a consequence of Proposition 2.1 which says that

$$e_0(\mathbb{M}) = h_0(\mathbb{M}) + h_1(\mathbb{M}) + \lambda(M_2/JM_1). \tag{2.17}$$

We are ready now to study the Hilbert function in the extremal case of Northcott's inequality and, at the same time, in the case of minimal multiplicity.

Theorem 2.9. *Let* $\mathbb{M} = \{M_j\}_{j \geq 0}$ *be a good* $\mathfrak{q}$-*filtration of the Cohen–Macaulay module M of dimension r and let J be an ideal generated by a maximal $\mathbb{M}$-superficial sequence for* $\mathfrak{q}$. *Let us consider the following conditions:*

(1) $s(\mathbb{M}) \leq 1$ *or, equivalently,* $P_{\mathbb{M}}(z) = \frac{h_0(\mathbb{M}) + h_1(\mathbb{M})z}{(1-z)^r}$.

(2) $e_1(\mathbb{M}) = h_1(\mathbb{M})$.

(3) $e_1(\mathbb{M}) = e_0(\mathbb{M}) - h_0(\mathbb{M})$.

(4) $e_0(\mathbb{M}) = h_0(\mathbb{M}) + h_1(\mathbb{M})$.

(5) $M_2 = JM_1$.

Then we have

$$\begin{array}{ccc}
1 & \Longrightarrow 2 & \Longrightarrow 4 \\
\Updownarrow & \Downarrow & \Updownarrow \\
1 & \Longleftarrow 3 & 5
\end{array}$$

If any of the first three equivalent conditions holds, then $gr_{\mathbb{M}}(M)$ *is Cohen–Macaulay.*

Proof. It is clear that $1 \Longrightarrow 2$, while, by using (2.17), we get $4 \Longleftrightarrow 5$. By looking at (2.16), it is clear that $e_1(\mathbb{M}) = h_1(\mathbb{M})$ implies $M_2 = JM_1$ and $e_1(\mathbb{M}) = e_0(\mathbb{M}) - h_0(\mathbb{M})$ so that $2 \Longrightarrow 3$ and 4. We need only to prove that $3 \Longrightarrow 1$.

If $e_1(\mathbb{M}) = e_0(\mathbb{M}) - h_0(\mathbb{M})$, then $M_2 \subseteq JM$, which implies $h_2(\mathbb{M}/JM) = 0$, and equality holds in (2.16). By Theorem 2.8, we get $gr_{\mathbb{M}}(M)$ is Cohen–Macaulay and $s(\mathbb{M}) \leq 2$. Hence $s(\mathbb{M}) = s(\mathbb{M}/JM) \leq 2$; but we have seen that $h_2(\mathbb{M}/JM) = 0$, hence $s(\mathbb{M}) = s(\mathbb{M}/JM) \leq 1$, as required. $\square$

Notice that the example given after Corollary 2.2 shows that in the above theorem the condition $M_2 = JM_1$ does not imply $s(\mathbb{M}) \leq 1$. In order to have this implication, we need to put some restriction on the filtration.

If L is any submodule of the given Cohen–Macaulay module M, and $\mathfrak{q}$ an $\mathfrak{m}$-primary ideal of A such that $\mathfrak{q}M \subseteq L$, let us consider the filtration

$$\mathbb{M}_L : \{M_0 = M, M_{j+1} = \mathfrak{q}^j L\} \tag{2.18}$$

for every $j \geq 0$. It is clear that $\mathbb{M}_L$ is a good $\mathfrak{q}$-filtration. $\mathbb{M}_L$ will be called the *filtration induced* by L and it will appear in most of the results from now on.

This filtration has the advantage that, by definition, one has $M_{j+1} = \mathfrak{q}M_j$ for every $j \geq 1$. For example this property allows us to conclude that the condition $M_2 = JM_1$ implies $s(\mathbb{M}) \leq 1$, see the below corollary. Notice that $\mathbb{M}_L$ is a generalization of the $\mathfrak{q}$-adic filtration $\mathbb{M} = \{\mathfrak{q}^n M\}$ considering $L = \mathfrak{q}M$.

In the application of the theory of filtered modules to the Fiber Cone, it will be useful to consider the filtration $\mathbb{M}_L$ in the particular case $L = \mathfrak{m}M$.

Corollary 2.6. *Let M be a given Cohen–Macaulay module, L a submodule and $\mathbb{M} = \mathbb{M}_L$ the good $\mathfrak{q}$-filtration on M induced by L. If J is an ideal generated by a maximal $\mathbb{M}$-superficial sequence for $\mathfrak{q}$, then all the conditions of the above theorem are equivalent.*

Proof. We prove that, for the filtration $\mathbb{M}$, we have $5 \Longrightarrow 1$. If $M_2 = JM_1$, then $\mathfrak{q}L = JL$ so that $M_{j+1} = JM_j$ for every $j \geq 1$. By Valabrega–Valla, this implies $gr_{\mathbb{M}}(M)$ is Cohen–Macaulay and $s(\mathbb{M}/JM) \leq 1$. Hence

$$s(\mathbb{M}) = s(\mathbb{M}/JM) \leq 1,$$

as wanted. $\square$

In the following theorem we study the equality $e_1(\mathbb{M}) = e_0(\mathbb{M}) - h_0(\mathbb{M}) + 1$. It turns out that we need some extra assumptions in order to get a complete description of this case. Nevertheless, the theorem extends results of Elias–Valla (see[26]), Guerrieri-Rossi (see [38]), Itoh (see [57]), Sally (see [92]) and Puthenpurakal (see [70]).

Theorem 2.10. *Let M be a Cohen–Macaulay module of dimension r, L be a submodule of M and $\mathbb{M} = \mathbb{M}_L$ the good $\mathfrak{q}$-filtration on M induced by L. If J is an ideal generated by a maximal $\mathbb{M}$-superficial sequence for $\mathfrak{q}$ and we assume $M_2 \cap JM = JM_1$, then the following conditions are equivalent.*

(1) $e_1(\mathbb{M}) = e_0(\mathbb{M}) - h_0(\mathbb{M}) + 1$.

(2) $e_1(\mathbb{M}) = h_1(\mathbb{M}) + 2$.

(3) $e_0(\mathbb{M}) = h_0(\mathbb{M}) + h_1(\mathbb{M}) + 1$ and $gr_{\mathbb{M}}(M)$ is Cohen–Macaulay.

(4) $P_{\mathbb{M}}(z) = \frac{h_0(\mathbb{M}) + h_1(\mathbb{M})z + z^2}{(1-z)^r}$.

Proof. It is clear that (4) implies (1). By (2.15), the assumption $M_2 \cap JM = JM_1$ gives the equality $h_1(\mathbb{M}) = h(\mathbb{M})$, hence, if $e_1(\mathbb{M}) = e_0(\mathbb{M}) - h_0(\mathbb{M}) + 1$, we get by (2.16)

$$
\begin{aligned}
e_1(\mathbb{M}) &= e_0(\mathbb{M}) - h_0(\mathbb{M}) + 1 \\
&\geq 2e_0(\mathbb{M}) - 2h_0(\mathbb{M}) - h_1(\mathbb{M}) \\
&= e_0(\mathbb{M}) - h_0(\mathbb{M}) + \lambda(M_2/JM_1) \\
&= h_1(\mathbb{M}) + 2\lambda(M_2/JM_1).
\end{aligned} \tag{2.19}
$$

Now $M_2 \neq JM_1$, otherwise, by Corollary 2.6, we have $e_1(\mathbb{M}) = e_0(\mathbb{M}) - h_0(\mathbb{M})$. Hence $\lambda(M_2/JM_1) = 1$ and we have equality above, so that $e_1(\mathbb{M}) = h_1(\mathbb{M}) + 2$. This proves that (1) implies (2).

Let us assume that $e_1(\mathbb{M}) = h_1(\mathbb{M}) + 2$. By Corollary 2.6, we have $M_2 \neq JM_1$, so that equality holds in (2.19) and $gr_{\mathbb{M}}(M)$ is Cohen–Macaulay by Theorem 2.8.

We need only to prove that (3) implies (4). Since $gr_{\mathbb{M}}(M)$ is Cohen–Macaulay, $\mathbb{M}$ and $\mathbb{M}/JM$ have the same h-polynomial so that

$$
P_{\mathbb{M}}(z) = \frac{h_{\mathbb{M}/JM}(z)}{(1-z)^r}
$$

But we have $e_0(\mathbb{M}) = e_0(\mathbb{M}/JM)$, $h_0(\mathbb{M}) = h_0(\mathbb{M}/JM)$ and also $h_1(\mathbb{M}) = h(\mathbb{M}) = h_1(\mathbb{M}/JM)$. Under the assumption $e_0(\mathbb{M}) = h_0(\mathbb{M}) + h_1(\mathbb{M}) + 1$ this implies

$$
\sum_{j \geq 2} h_j(\mathbb{M}/JM) = 1.
$$

Now if $h_2(\mathbb{M}/JM) = 0$, then $\mathfrak{q}L \subseteq \mathfrak{q}^2 L + JM$ which implies $h_t(\mathbb{M}/JM) = 0$ for every $t \geq 2$. Hence $h_2(\mathbb{M}/JM) \neq 0$ and since $h_j(\mathbb{M}/JM) \geq 0$ for every j, we get $h_2(\mathbb{M}/JM) = 1$ and $h_j(\mathbb{M}/JM) = 0$ for $j \geq 3$. The proof of the theorem is now complete. $\qquad\square$

As already shown in [38], the assumption $M_2 \cap JM = JM_1$ is essential. The Cohen–Macaulay local ring $A = k[[t^4, t^5, t^6, t^7]]$ and the primary ideal $\mathfrak{q} = (t^4, t^5, t^6)$ give, with $M = A$ and $L = \mathfrak{q}$, an example where $e_1(\mathbb{M}) = e_0(\mathbb{M}) - h_0(\mathbb{M}) + 1$ but $gr_{\mathbb{M}}(M)$ is not Cohen–Macaulay.

As in [92] we remark that we can apply the theorem when $\mathbb{M}$ is the $\mathfrak{q}$-adic filtration and $\mathfrak{q}$ is integrally closed. Namely Itoh has shown that if $\mathfrak{q}$ is any $\mathfrak{m}$-primary ideal, then

$$
J \cap \overline{\mathfrak{q}^2} = J\overline{\mathfrak{q}}.
$$

Hence, if $\mathfrak{q}$ is integrally closed (e.g. $\mathfrak{q} = \mathfrak{m}$), then

$$
\mathfrak{q}^2 \cap J \subseteq J \cap \overline{\mathfrak{q}^2} = J\overline{\mathfrak{q}} = J\mathfrak{q},
$$

so $J\mathfrak{q} = \mathfrak{q}^2 \cap J$.

In order to get rid of the assumption $M_2 \cap JM = JM_1$, we need one more ingredient, the study of the Ratliff–Rush filtration.

In the next chapter, after discussing the basic properties of this filtration, we apply the results concerning the second Hilbert coefficient $e_2(\mathbb{M})$ thus completing the study of the equality

$$e_1(\mathbb{M}) = e_0(\mathbb{M}) - h_0(\mathbb{M}) + 1$$

(see Theorem 3.3).

Chapter 3
Bounds for $e_2(\mathbb{M})$

If M is a Cohen–Macaulay module, then in the previous chapter we showed that $e_0(\mathbb{M})$ and $e_1(\mathbb{M})$ are positive integers.

In the classical case of an $\mathfrak{m}$-primary ideal $\mathfrak{q}$ of a Cohen–Macaulay local ring A, as far as the higher Hilbert coefficients are concerned, it is a famous result of M. Narita that $e_2(\mathfrak{q}) \geq 0$ [63]. This result is extended here to the case of modules. In the same paper, Narita also showed that if $\dim A = 2$, then $e_2(\mathfrak{q}) = 0$ if and only if $\mathfrak{q}^n$ has reduction number one for large n. Consequently, $gr_{\mathfrak{q}^n}(A)$ is Cohen–Macaulay. There are examples which show that the result cannot be extended to higher dimensions. Very recently Puthenpurakal presented some new results concerning this problem, see [71]. Interesting results on $e_2(\mathfrak{q})$ can also be found in [14] which investigates the interplay between the integrality, or even the normality, of the ideal $\mathfrak{q}$ and $e_2(\mathfrak{q})$. Classical bounds for $e_2(\mathbb{M})$ can be improved and reformulated in our general setting by using special good $\mathfrak{q}$-filtrations on the module M. We shall introduce the notion of the Ratliff–Rush filtration for studying Hilbert coefficients. This is a device which will be crucial also in the next chapter.

Unfortunately, the positivity of $e_2(\mathbb{M})$ does not extend to the higher Hilbert coefficients. Indeed, in [63] M. Narita showed that $e_3(\mathbb{M})$ can be negative. However, a remarkable result of S. Itoh says that if $\mathfrak{q}$ is a normal ideal then $e_3(\mathfrak{q}) \geq 0$ [58]. A nice proof of this result was also given by S. Huckaba and C. Huneke in [51]. In general, it seems that the integral closedness (or the normality) of the ideal $\mathfrak{q}$ has non trivial consequences for the Hilbert coefficients of $\mathfrak{q}$ and, ultimately, for depth $gr_{\mathfrak{q}}(A)$.

3.1 The Ratliff–Rush filtration

Given a good $\mathfrak{q}$-filtration on the module M, we shall introduce a new filtration which was constructed by Ratliff and Rush in [73]. Here we extend the construction to the general case of a filtered module by following the definition given by W. Heinzer et al. in [45, Sect. 6]. A further generalization was studied by T.J. Puthenpurakal and F. Zulfeqarr in [72].

M.E. Rossi and G. Valla, *Hilbert Functions of Filtered Modules*, Lecture Notes
of the Unione Matematica Italiana 9, DOI 10.1007/978-3-642-14240-6_3,
© Springer-Verlag Berlin Heidelberg 2010

Let $\mathfrak{q}$ be an $\mathfrak{m}$-primary ideal in A and let $\mathbb{M}$ be a good $\mathfrak{q}$-filtration on the module M. We define the filtration $\widetilde{\mathbb{M}}$ on M by letting

$$\widetilde{M}_n := \bigcup_{k \geq 1} (M_{n+k} :_M \mathfrak{q}^k).$$

If there is no confusion, we will omit the subscript M in the colon. It is clear that $\widetilde{M}_0 = \widetilde{M} = M$ and, for every $n \geq 0$, $M_n \subseteq \widetilde{M}_n$.

Further, since M is Noetherian, there is a positive integer t, depending on n such that

$$\widetilde{M}_n = M_{n+k} : \mathfrak{q}^k \quad \forall k \geq t.$$

The filtration $\widetilde{\mathbb{M}}$ is called the *Ratliff–Rush* filtration associated to $\mathbb{M}$.

If $\mathbb{M}$ is the $\mathfrak{q}$-adic filtration of $M = A$, then for every integer n there exists an integer k such that

$$\widetilde{M}_n = \widetilde{\mathfrak{q}^n} = \mathfrak{q}^{n+k} : \mathfrak{q}^k.$$

The most important properties of $\widetilde{\mathbb{M}}$ are collected in the following lemma.

Lemma 3.1. *Let $\mathbb{M}$ be a good $\mathfrak{q}$-filtration on the r-dimensional module M, such that $\operatorname{depth}_{\mathfrak{q}}(M) \geq 1$. Then we have:*

(1) There exists an integer n_0 such that $M_n = \widetilde{M}_n$ for all $n \geq n_0$.

(2) $\widetilde{\mathbb{M}}$ is a good $\mathfrak{q}$-filtration on M.

(3) If a is $\mathbb{M}$-superficial for $\mathfrak{q}$, then it is also $\widetilde{\mathbb{M}}$-superficial for $\mathfrak{q}$.

(4) $\widetilde{\mathbb{M}}$ and $\mathbb{M}$ share the same Hilbert–Samuel polynomial so that $e_i(\widetilde{\mathbb{M}}) = e_i(\mathbb{M})$ for every $i = 0, \ldots, r$.

(5) If a is $\mathbb{M}$-superficial for $\mathfrak{q}$, then $\widetilde{M}_{j+1} : a = \widetilde{M}_j$ for every $j \geq 0$, so that $\operatorname{depth} gr_{\widetilde{\mathbb{M}}}(M) \geq 1$.

(6) $\operatorname{depth} gr_{\mathbb{M}}(M) \geq 1$ if and only if $M_n = \widetilde{M}_n$ for every n.

Proof. Let a be $\mathbb{M}$-superficial for $\mathfrak{q}$. Since $\operatorname{depth}_{\mathfrak{q}}(M) \geq 1$, a is a regular element for $\mathfrak{m}$ and, by Theorem 1.2, there exists an integer n such that $M_{j+1} : a = M_j$ for every $j \geq n$. We have

$$\widetilde{M}_n = M_{n+k} : \mathfrak{q}^k \subseteq M_{n+k} : a^k = (M_{n+k} : a) :_M a^{k-1}$$
$$= M_{n+k-1} : a^{k-1} = \cdots = M_{n+1} : a = M_n.$$

which proves the first assertion. Also we have

$$\mathfrak{q}\widetilde{M}_n = \mathfrak{q}(M_{n+k} : \mathfrak{q}^k) \subseteq \mathfrak{q}M_{n+k} : \mathfrak{q}^k \subseteq M_{n+k+1} : \mathfrak{q}^k \subseteq \widetilde{M}_{n+1}$$

which proves that $\widetilde{\mathbb{M}}$ is a $\mathfrak{q}$-filtration. Further, since $M_n = \widetilde{M}_n$ for $n \gg 0$, $\widetilde{\mathbb{M}}$ is $\mathfrak{q}$-good, a is $\widetilde{\mathbb{M}}$-superficial for $\mathfrak{q}$ and $\widetilde{\mathbb{M}}$ and $\mathbb{M}$ share the same Hilbert–Samuel polynomial. This proves that $e_i(\widetilde{\mathbb{M}}) = e_i(\mathbb{M})$ for every $i = 0, \ldots, r$.

We prove now that $\widetilde{M}_{j+1} : a = \widetilde{M}_j$ for every $j \geq 0$. It is clear that we can find an integer k such that $M_{j+1+k} : a = M_{j+k}$ and $\widetilde{M}_{j+1} = M_{j+1+k} : \mathfrak{q}^k$. Then we get

$$\widetilde{M}_{j+1} : a = (M_{j+1+k} : \mathfrak{q}^k) : a = (M_{j+1+k} : a) : \mathfrak{q}^k = M_{j+k} : \mathfrak{q}^k \subseteq \widetilde{M}_j.$$

Finally we must prove that depth $gr_{\mathbb{M}}(M) \geq 1$ implies $\widetilde{M}_n = M_n$ for every n. But we have

$$\widetilde{M}_n = M_{n+k} : \mathfrak{q}^k \subseteq M_{n+k} : a^k = M_n$$

because a^* is a regular element on $gr_{\mathbb{M}}(M)$. □

The filtration $\widetilde{\mathbb{M}}$ on M is a good $\mathfrak{q}$-filtration because, by the definition, we have $\mathfrak{q}\widetilde{M}_n \subseteq \widetilde{M}_{n+1}$ and, by Lemma 3.1, for large n we have $\widetilde{M}_{n+1} = M_{n+1} = \mathfrak{q}M_n = \mathfrak{q}\widetilde{M}_n$.

It's worth recalling that T.J. Puthenpurakal and F. Zulfeqarr in [72] and in [71] further analyzed the case when $\mathfrak{q}$ is not a regular ideal, i.e. it does not contain an M-regular element.

3.2 Bounds for $e_2(\mathbb{M})$

The Ratliff–Rush filtration is a very useful tool for proving results about Hilbert coefficients, but in general it does not behave well in inductive arguments, except in few cases, for instance for the integrally closed ideals.

Denote by $\overline{\mathfrak{q}}$ the integral closure of the ideal $\mathfrak{q}$ in A. It is easy to see that

$$\mathfrak{q} \subseteq \widetilde{\mathfrak{q}} \subset \overline{\mathfrak{q}}.$$

Hence, if $\mathfrak{q}$ is an integrally closed ideal, then $\widetilde{\mathfrak{q}} = \mathfrak{q}$ (say that $\mathfrak{q}$ is Ratliff–Rush closed). In this case the Ratliff–Rush closure commutes with the quotient by a superficial element. In fact S. Itoh proved that if $\mathfrak{q}$ is an integrally closed ideal, there exists a superficial element $a \in \mathfrak{q}$ such that $\mathfrak{q}B$ is integrally closed in $B = A/aA$, hence $\mathfrak{q}B$ is Ratliff–Rush closed in B. In particular

$$\widetilde{\mathfrak{q}}B = \widetilde{\mathfrak{q}B}.$$

This is not true in general.

Superficial elements do not behave well, even if we consider Ratliff–Rush closed ideals (non integrally closed). Consider $\mathfrak{q} = (x^l, xy^{l-1}, y^l), l > 2$, in $A = k[[x,y]]$ (see [78]); in this case all the powers of $\mathfrak{q}$ are Ratliff–Rush closed, nevertheless there is no superficial element $a \in \mathfrak{q}$ for which $\mathfrak{q}/(a)$ is not Ratliff–Rush closed in $B = A/aA$, hence $\widetilde{\mathfrak{q}}B \neq \widetilde{\mathfrak{q}B}$. In [71, Theorems 3.3 and 5.5], T. Puthenpurakal gives a complete characterization of the existence of a superficial element $a \in \mathfrak{q}$ for which

$$\widetilde{\mathfrak{q}^i}B = \widetilde{\mathfrak{q}^iB}$$

for every integer i.

An important fact proved by Huckaba and Marley in [52, Corollary 4.13] is an easy consequence of our approach.

Let us assume $\mathbb{M}$ is a good q-filtration on the two-dimensional Cohen–Macaulay module M, then by Theorem 2.5(c) and Lemma 3.1, (4) and (5) we have

$$e_1(\mathbb{M}) = e_1(\widetilde{\mathbb{M}}) = \sum_{j\geq 0} v_j(\widetilde{\mathbb{M}}) \quad e_2(\mathbb{M}) = e_2(\widetilde{\mathbb{M}}) = \sum_{j\geq 1} jv_j(\widetilde{\mathbb{M}}) \qquad (3.1)$$

We recall that $v_j(\widetilde{\mathbb{M}}) = \lambda(\widetilde{M}_{j+1}/J\widetilde{M}_j)$ where J is a maximal $\mathbb{M}$-superficial sequence for q, hence by Lemma 3.1 (3), a maximal $\widetilde{\mathbb{M}}$-superficial sequence for q. As a first application of the previous formula we obtain a short proof of the non negativity of $e_2(\mathbb{M})$.

Proposition 3.1. *Let $\mathbb{M}$ be a good q-filtration of the Cohen–Macaulay module M of dimension r. Then*

$$e_2(\mathbb{M}) \geq 0.$$

Proof. If $r = 1$, it is clear by Lemma 2.2. Let $r \geq 2$, by Proposition 1.2 we may assume $r = 2$. Hence by (3.1)

$$e_2(\mathbb{M}) = e_2(\widetilde{\mathbb{M}}) = \sum_{j\geq 1} jv_j(\widetilde{\mathbb{M}}) \geq 0.$$

□

The following example given in [14] shows that $e_2(q) = 0$ does not imply the Cohen–Macaulayness of $gr_q(A)$.

Example 3.1. Let A be the regular local ring $k[\![x,y,z]\!]$, with k a field and x,y,z indeterminates and consider $q = (x^2 - y^2, y^2 - z^2, xy, xz, yz)$, then

$$P_q(z) = \frac{5 + 6z^2 - 4z^3 + z^4}{(1-z)^3}.$$

In particular, $e_2(q) = 0$ and we prove that $gr_q(A)$ has depth zero. In fact we can find a superficial element for q whose initial form is not regular on $gr_q(A)$. Computing with CoCoA the Hilbert coefficients $e_i(q/(xy))$ we can see that xy is a superficial element for q (see Remark 1.2), but its initial form is a zero-divisor on $gr_q(A)$ since $P_q(z) \neq P^1_{q/(xy)}(z)$ (see Proposition 1.2).

In the two-dimensional case, one can prove that if $e_2(\mathbb{M}) = 0$, then $gr_{\widetilde{\mathbb{M}}}(M)$ is Cohen–Macaulay. We prove this in the next theorem, where we also extend results by Sally and Narita (see [93] and [63]). An extension to dimension $r \geq 2$ was proved by Puthenpurakal in [71].

Theorem 3.1. *Let $\mathbb{M} = \{M_j\}_{j\geq 0}$ be a good q-filtration of the Cohen–Macaulay module M of dimension 2 and let J be an ideal generated by a maximal $\mathbb{M}$-superficial sequence for q. Then:*

(1) $e_2(\mathbb{M}) \geq e_1(\mathbb{M}) - e_0(\mathbb{M}) + \lambda(M/\widetilde{M}_1) \geq 0$.

(2) If $e_2(\mathbb{M}) = 0$ and $M_1 = \widetilde{M}_1$, then $e_1(\mathbb{M}) = e_0(\mathbb{M}) - h_0(\mathbb{M})$ so that $s(\mathbb{M}) \leq 1$ and $gr_{\mathbb{M}}(M)$ is Cohen–Macaulay.

(3) $gr_{\widetilde{\mathbb{M}}}(M)$ is Cohen–Macaulay if at least one of the following conditions holds:

    (a) $e_2(\mathbb{M}) = 0$,

    (b) $e_2(\mathbb{M}) = e_1(\mathbb{M}) - e_0(\mathbb{M}) + \lambda(M/\widetilde{M}_1)$ and $\widetilde{M}_2 \cap JM = J\widetilde{M}_1$.

Proof. Since depth $gr_{\widetilde{\mathbb{M}}}(M) \geq 1 = r - 1$, we have $e_2(\mathbb{M}) = e_2(\widetilde{\mathbb{M}}) = \sum_{j \geq 1} j v_j(\widetilde{\mathbb{M}})$ and $e_1(\mathbb{M}) = e_1(\widetilde{\mathbb{M}}) = \sum_{j \geq 0} v_j(\widetilde{\mathbb{M}})$. Hence we get

$$
\begin{aligned}
e_2(\mathbb{M}) &= e_1(\mathbb{M}) - v_0(\widetilde{\mathbb{M}}) + \sum_{j \geq 2}(j-1)v_j(\widetilde{\mathbb{M}}) \\
&= e_1(\mathbb{M}) - \lambda(\widetilde{M}_1/JM) + \sum_{j \geq 2}(j-1)v_j(\widetilde{\mathbb{M}}) \\
&= e_1(\mathbb{M}) - e_0(\mathbb{M}) + \lambda(M/\widetilde{M}_1) + \sum_{j \geq 2}(j-1)v_j(\widetilde{\mathbb{M}}) \\
&\geq e_1(\mathbb{M}) - e_0(\mathbb{M}) + \lambda(M/\widetilde{M}_1) \geq 0
\end{aligned}
$$

where the last inequality follows by (2.16) applied to the Ratliff–Rush filtration $\widetilde{M}$. This proves (1) which trivially gives (2). As for (3), if $e_2(\mathbb{M}) = 0$, then $e_1(\mathbb{M}) = e_0(\mathbb{M}) - \lambda(M/\widetilde{M}_1)$ and $gr_{\widetilde{\mathbb{M}}}(M)$ is Cohen–Macaulay by Theorem 2.9.

If $e_2(\mathbb{M}) = e_1(\mathbb{M}) - e_0(\mathbb{M}) + \lambda(M/\widetilde{M}_1)$, then $v_j(\widetilde{\mathbb{M}}) = 0$ for every $j \geq 2$. This means that $\widetilde{M}_{j+1} = J\widetilde{M}_j$ for every $j \geq 2$, and since $\widetilde{M}_2 \cap JM = J\widetilde{M}_1$, $gr_{\widetilde{\mathbb{M}}}(M)$ is Cohen–Macaulay by Valabrega–Valla. $\qquad\square$

The inequality $e_2(\mathbb{M}) \geq e_1(\mathbb{M}) - e_0(\mathbb{M}) + \lambda(M/\widetilde{M}_1) \geq 0$ was proved by Sally in [93, Corollary 2.5] in the special case $M_j = \mathfrak{q}^j$. The methods there involve the local cohomology of the Rees ring.

As a consequence of the previous theorem we obtain a classical result by Narita (see [63]).

Corollary 3.1. *Let $\mathfrak{q}$ be an $\mathfrak{m}$-primary ideal which is integrally closed in a Cohen–Macaulay local ring $(A, \mathfrak{m})$ of dimension $r \geq 2$. Then:*

(1) $e_2(\mathfrak{q}) \geq e_1(\mathfrak{q}) - e_0(\mathfrak{q}) + \lambda(A/\mathfrak{q})$

(2) *If $e_2(\mathfrak{q}) = 0$ then $gr_{\mathfrak{q}}(A)$ is Cohen–Macaulay and $e_i(\mathfrak{q}) = 0$ for $i \geq 2$.*

In the above corollary we get rid of the assumption $\dim(M) = 2$ of Theorem 3.1 since $\mathfrak{q}$ is integrally closed and, by Proposition 1.1, we can find a superficial sequence $a_1, \ldots, a_{r-2}$ in $\mathfrak{q}$ such that $\mathfrak{q}/(a_1, \ldots, a_{r-2})$ is integrally closed in $A/(a_1, \ldots, a_{r-2})$ which is a local Cohen–Macaulay ring of dimension two. Moreover, by Proposition 1.2, the numerical invariants involved are preserved going modulo

$(a_1, \ldots, a_{r-2})$ and we may apply Theorem 3.1 in order to prove (1). The second assertion comes from (1) and Theorem 2.9.

Example 3.1 shows that in Corollary 3.1 the assumption that the ideal $\mathfrak{q}$ is integrally closed cannot be weakened. In that example $e_2(\mathfrak{q}) = 0$, but $gr_\mathfrak{q}(A)$ is not Cohen–Macaulay.

Notice that inequality (1) was already proved by Itoh in [58]. Assertion 2 of Corollary 3.1 was proved by Puthenpurakal in [70].

Later, it was conjectured by Valla [107, (6.20)] that if the equality $e_2 = e_1 - e_0 + \lambda(A/\mathfrak{q})$ holds when $\mathfrak{q}$ is the maximal ideal $\mathfrak{m}$ of A, then the associated graded ring $gr_\mathfrak{m}(A)$ is Cohen–Macaulay. Unfortunately, the following example given by Wang shows that the conjecture is false.

Example 3.2. Let A be the two-dimensional local Cohen–Macaulay ring

$$k[\![x, y, t, u, v]\!]/(t^2, tu, tv, uv, yt - u^3, xt - v^3),$$

with k a field and x, y, t, u, v indeterminates. Let $\mathfrak{m}$ be the maximal ideal of A. One has that the associated graded ring $gr_\mathfrak{m}(A)$ has depth zero and

$$P_\mathfrak{m}(z) = \frac{1 + 3z + 3z^3 - z^4}{(1-z)^2}.$$

In particular, one has $e_2 = e_1 - e_0 + 1$, that is, e_2 is minimal according to the bound proved in Corollary 3.1.

Hence the condition $\lambda(A/\mathfrak{q}) = e_0(\mathfrak{q}) - e_1(\mathfrak{q}) + e_2(\mathfrak{q})$ does not imply that $gr_\mathfrak{q}(A)$ is Cohen–Macaulay even for an integrally closed ideal $\mathfrak{q}$. However, Corso, Polini and Rossi in [14] proved that the conjecture is true if $\mathfrak{q}$ is normal (i.e. $\mathfrak{q}^n$ is integrally closed for every n).

Theorem 3.2. *Let $\mathfrak{q}$ be a normal $\mathfrak{m}$-primary ideal in a Cohen–Macaulay local ring $(A, \mathfrak{m})$ of dimension $r \geq 2$. Then:*

(1) $e_2(\mathfrak{q}) \geq e_1(\mathfrak{q}) - e_0(\mathfrak{q}) + \lambda(A/\mathfrak{q})$.

(2) If $e_2(\mathfrak{q}) = e_1(\mathfrak{q}) - e_0(\mathfrak{q}) + \lambda(A/\mathfrak{q})$ then $gr_\mathfrak{q}(A)$ is Cohen–Macaulay.

In the above result (1) follows by Corollary 3.1. The crucial point in the proof of (2) is to prove that the reduction number of $\mathfrak{q}$ is at most two. Here a result by Itoh in [57] was fundamental. Another context where the normality plays an important role in the study of the reduction number of $\mathfrak{q}$ is considered by Lipman and Teissier [61, Corollary 5.4] when we consider pseudo-rational two-dimensional normal local ring.

We present here a short proof of a result of Narita for modules which characterizes $e_2(\mathbb{M}) = 0$ when M is a Cohen–Macaulay module of dimension two and $\mathbb{M}$

is the q-adic filtration on M. We will write $e_i(\mathfrak{q}^n M)$ when we consider the Hilbert coefficients of the $\mathfrak{q}^n$-adic filtration on M with n a fixed integer.

Proposition 3.2. *Let* $\mathfrak{q}$ *be an* $\mathfrak{m}$-*primary ideal and let* M *be a Cohen–Macaulay module of dimension two.*

Then $e_2(\mathfrak{q}M) = 0$ *if and only if* $\mathfrak{q}^n M$ *has reduction number one for some positive integer* n. *Under these circumstances* $gr_{\mathfrak{q}^n M}(M)$ *is Cohen–Macaulay.*

Proof. We first recall that $e_2(\mathfrak{q}M) = e_2(\mathfrak{q}^m M)$ for every positive integer m. Assume $e_2(\mathfrak{q}M) = 0$ and let n be an integer such that $\widetilde{\mathfrak{q}^n M} = \mathfrak{q}^n M$. Hence by Theorem 3.1,
$$0 = e_2(\mathfrak{q}^n M) \geq e_1(\mathfrak{q}^n M) - e_0(\mathfrak{q}^n M) + \lambda(M/\widetilde{\mathfrak{q}^n M}) = e_1(\mathfrak{q}^n M) - e_0(\mathfrak{q}^n M)$$
$+\lambda(M/\mathfrak{q}^n M)$. Hence $e_1(\mathfrak{q}^n M) - e_0(\mathfrak{q}^n M) + \lambda(M/\mathfrak{q}^n M) = 0$ because it cannot be negative by Northcott's inequality. The result follows now by Theorem 2.9. For the converse, if $\mathfrak{q}^n M$ has reduction number one for some n, then $e_2(\mathfrak{q}^n M) = 0$ and $gr_{\mathfrak{q}^n M}(M)$ is Cohen–Macaulay again by Theorem 2.9. In particular $e_2(\mathfrak{q}M) = e_2(\mathfrak{q}^n M) = 0$. $\qquad\square$

We remark that Narita's result cannot be extended to a local Cohen–Macaulay ring of dimension >2 without changing the statement. The ideal $\mathfrak{q}$ presented in Example 3.1 satisfies $e_2(\mathfrak{q}) = 0$, however $\mathfrak{q}^n$ has reduction number greater than one for every n. In fact, it is enough to remark that $\mathfrak{q}$ does not have reduction number one ($gr_{\mathfrak{q}}(A)$ is not Cohen–Macaulay) and $\mathfrak{q}^n = (x, y, z)^{2n}$ for $n > 1$ which has reduction number two. An extension of Narita's result to the higher dimensional case is given by Puthenpurakal in [71].

We can prove now the following result which is, both, an extension and a completion of a deep theorem proved by Sally in [92]. This result is new even in the classical case and it completes the study of the equality $e_1(\mathbb{M}) = e_0(\mathbb{M}) - h_0(\mathbb{M}) + 1$ (see Chap. 2).

Theorem 3.3. *Let* M *be a Cohen–Macaulay module of dimension* $r \geq 2$, L *a submodule of* M *and* $\mathbb{M} = \mathbb{M}_L$ *the good* $\mathfrak{q}$-*filtration on* M *induced by* L. *Assume that* $e_1(\mathbb{M}) = e_0(\mathbb{M}) - h_0(\mathbb{M}) + 1$. *Then the following conditions are equivalent:*

(1) $e_2(\mathbb{M}) \neq 0$.
(2) $e_2(\mathbb{M}) = 1$.
(3) depth $gr_{\mathbb{M}}(M) \geq r - 1$.
(4) $P_{\mathbb{M}}(z) = \frac{h_0(\mathbb{M}) + h_1(\mathbb{M})z + z^2}{(1-z)^r}$.

Proof. We recall that we are considering the filtration

$$\mathbb{M}_L : \quad M \supseteq L \supseteq \mathfrak{q}L \supseteq \cdots \supseteq \mathfrak{q}^j L \supseteq \ldots$$

First we prove that (1), (2) and (3) are equivalent. As usual, J is an ideal generated by a maximal sequence of $\mathbb{M}$-superficial elements for q. Let us first consider the case $r = 2$. We have

$$v_0(\widetilde{\mathbb{M}}) = \lambda(\widetilde{M}_1/JM) = e_0(\mathbb{M}) - h_0(\mathbb{M}) + \lambda(\widetilde{M}_1/M_1).$$

and depth $gr_{\widetilde{\mathbb{M}}}(M) \geq 1 = r - 1$. This implies

$$e_2(\mathbb{M}) = e_2(\widetilde{\mathbb{M}}) = \sum_{j \geq 1} j v_j(\widetilde{\mathbb{M}})$$

$$e_0(\mathbb{M}) - h_0(\mathbb{M}) + 1 = e_1(\mathbb{M}) = e_1(\widetilde{\mathbb{M}}) = \sum_{j \geq 0} v_j(\widetilde{\mathbb{M}})$$

so that

$$\sum_{j \geq 1} v_j(\widetilde{\mathbb{M}}) = \sum_{j \geq 0} v_j(\widetilde{\mathbb{M}}) - v_0(\widetilde{\mathbb{M}}) = 1 - \lambda(\widetilde{M}_1/M_1). \tag{3.2}$$

Let us assume that (1) holds, then $\sum_{j \geq 1} v_j(\widetilde{\mathbb{M}}) > 0$, so that $\sum_{j \geq 1} v_j(\widetilde{\mathbb{M}}) = 1$ and $\lambda(\widetilde{M}_1/M_1) = 0$. But if $\widetilde{M}_1 = M_1$, we cannot have $v_1(\widetilde{\mathbb{M}}) = 0$, otherwise

$$M_2 \subseteq \widetilde{M}_2 = J\widetilde{M}_1 = JM_1,$$

and, by Theorem 2.9, $e_1(\mathbb{M}) = e_0(\mathbb{M}) - h_0(\mathbb{M})$, a contradiction. Hence $v_1(\widetilde{\mathbb{M}}) = 1$ and $v_j(\widetilde{\mathbb{M}}) = 0$ for every $j \geq 2$, which implies $e_2(\mathbb{M}) = 1$. This proves that (1) implies (2).

Let now assume that $e_2(\mathbb{M}) = 1$. Then we must have $v_1(\widetilde{\mathbb{M}}) = 1$ and $v_j(\widetilde{\mathbb{M}}) = 0$ for every $j \geq 2$, so that, by (3.2), $\widetilde{M}_1 = M_1$. Thus

$$1 = \lambda(\widetilde{M}_2/JM_1) \geq \lambda(M_2/JM_1) \geq 1$$

which implies $\widetilde{M}_2 = M_2$. Now if $j \geq 2$ and $\widetilde{M}_j = M_j$, then we have

$$M_{j+1} \subseteq \widetilde{M}_{j+1} = J\widetilde{M}_j = JM_j \subseteq M_{j+1}.$$

Hence, by induction, we get $\widetilde{M}_t = M_t$ for every $t \geq 1$. By the above Lemma, this implies depth $gr_{\mathbb{M}}(M) > 0$, thus proving that (2) implies (3).

Finally, the condition depth $gr_{\mathbb{M}}(M) > 0$ implies $\widetilde{M}_1 = M_1$, hence $\sum_{j \geq 1} v_j(\widetilde{\mathbb{M}}) = 1$ and $e_2(\mathbb{M}) \neq 0$. This completes the proof of the equivalence of (1), (2) and (3) in the case $r = 2$.

Let us now consider the general case, when $r \geq 3$. Let $\mathfrak{a}$ be an ideal generated by an $\mathbb{M}$-superficial sequence for q of length $r - 2$. Then we have $e_i(\mathbb{M}) = e_i(\mathbb{M}/\mathfrak{a}M)$ for $i = 0, 1, 2$ and $h_0(\mathbb{M}) = h_0(\mathbb{M}/\mathfrak{a}M)$. Hence the assumption holds for the two-dimensional Cohen–Macaulay module $M/\mathfrak{a}M$. The conclusion follows because, by Sally's machine, depth $gr_{\mathbb{M}}(M) \geq r - 1$ if and only if depth $gr_{\mathbb{M}}(M/\mathfrak{a}M) \geq 1$.

We end the proof of the theorem by proving that (2) is equivalent to (4). We notice that if $e_2(\mathbb{M}) = 1$ and depth $gr_{\mathbb{M}}(M) \geq r-1$, then, by Theorem 2.5, we get

$$1 = e_2(\mathbb{M}) = \sum_{j \geq 1} j v_j(\mathbb{M}).$$

Hence $v_1(\mathbb{M}) = 1$ and $v_j(\mathbb{M}) = 0$ for $j \geq 2$. Since $e_i(\mathbb{M}) = \sum_{j \geq i-1} \binom{j}{i-1} v_j(\mathbb{M})$, we also get $e_j(\mathbb{M}) = 0$ for $j \geq 3$. These values of the e_i give the required Hilbert series and from the Hilbert series we can compute the e_i. □

The following example from [93] shows that in the above result the assumption $e_2(\mathbb{M}) \neq 0$ is essential. We remark that the q-adic filtration of A is a filtration of the type $\mathbb{M}_L$ induced on A by $L = \mathfrak{q}$ itself.

Example 3.3. Consider the ideal $q = (x^4, x^3y, xy^3, y^4) \subseteq A = k[[x,y]]$. The ideal $\mathfrak{q}$ is not integrally closed and if we consider on A the q-adic filtration, we have

$$P_{\mathfrak{q}}(z) = \frac{11 + 3z + 3z^2 - z^3}{(1-z)^2}.$$

This gives $e_0 = 16$, $e_1 = 6$, $e_2 = 0$, $h_0(\mathfrak{q}) = 11$, so that $e_1(\mathfrak{q}) = e_0(\mathfrak{q}) - h_0(\mathfrak{q}) + 1$. It is clear that $x^2y^2 \notin \mathfrak{q}$ while $x^2y^2\mathfrak{q} \subseteq \mathfrak{q}^2$ so that $\tilde{\mathfrak{q}} \neq \mathfrak{q}$ and $gr_{\mathfrak{q}}(A)$ has depth zero by Lemma 3.1.

Very little is known about the Hilbert Function of the filtered module M when $e_2(\mathbb{M}) = 0$ and $M_1 \neq \tilde{M}_1$. Later we shall give more information, provided $e_1(\mathbb{M}) = e_0(\mathbb{M}) - h_0(\mathbb{M}) + 1$.

The following example shows that in the above theorem the assumption $e_1(\mathbb{M}) = e_0(\mathbb{M}) - h_0(\mathbb{M}) + 1$ and $e_2(\mathbb{M}) = 1$ does not imply $gr_{\mathbb{M}}(A)$ is Cohen–Macaulay.

Example 3.4. Consider the ideal $q = (x^6, x^5y^3, x^4y^7, x^3y^8, x^2y^{10}, xy^{11}, y^{22})$ in $A = k[[x,y]]$. We have

$$P_{\mathfrak{q}}(z) = \frac{61 + 26z + z^2}{(1-z)^2}.$$

This gives $e_0(\mathfrak{q}) = 88$, $e_1(\mathfrak{q}) = 28$, $e_2(\mathfrak{q}) = 1$, $h_0(\mathfrak{q}) = 61$, so that we have $e_1(\mathfrak{q}) = e_0(\mathfrak{q}) - h_0(\mathfrak{q}) + 1$. However $gr_{\mathfrak{q}}(A)$ is not Cohen–Macaulay because $\mathfrak{q}$ is an m-primary ideal in a regular ring of dimension two and $s(\mathfrak{q}) = 2 > 1$ (see [52, Theorem A and Proposition 2.6], [9, Proposition 2.9]).

The example above underlines the difference between the case of the m-adic filtration and the more general case of the filtration induced by the powers of an m-primary ideal q. In the first case [26] Elias and Valla proved that if the degree of the h-polynomial is less than or equal to two, then the associated graded ring is

Cohen–Macaulay. The above example shows that this is not the case when $\mathfrak{q}$ is not maximal, even if A is regular and $h_2 = 1$.

The second statement in Theorem 2 of [70] says that if $M = A$ is Cohen–Macaulay, $\mathfrak{q} = \mathfrak{m}$, $\dim A = 2$ and $e_1(A) = 2e_0(A) - \mu(\mathfrak{m}) + 1$, then either $gr_{\mathfrak{m}^n}(A)$ is Cohen–Macaulay for $n \gg 0$ or depth $gr_{\mathfrak{m}}(A) \geq 1$.

We notice that we have

$$2e_0(A) - \mu(\mathfrak{m}) + 1 = 2e_0(A) - 2h_0(A) - h(A) + 1,$$

so that this last result will be a consequence of the following theorem which is a step further after Theorem 2.8.

Theorem 3.4. *Let M be a Cohen–Macaulay module of dimension $r \geq 2$, L a submodule of M and $\mathbb{M} = \mathbb{M}_L$ the good $\mathfrak{q}$-filtration on M induced by L. Let J be generated by a maximal $\mathbb{M}$-superficial sequence for $\mathfrak{q}$ and assume that $e_1(\mathbb{M}) = = 2e_0(\mathbb{M}) - 2h_0(\mathbb{M}) - h(\mathbb{M}) + 1$, $\widetilde{M}_1 = M_1$ and $\widetilde{M}_2 \cap JM = JM_1$. Then we have:*

(1) If $\widetilde{M}_2 \neq M_2$, then $gr_{\widetilde{\mathbb{M}}}(M)$ is Cohen–Macaulay.

(2) If $r = 2$, then

  (a) $e_2(\mathbb{M}) = e_0(\mathbb{M}) - h_0(\mathbb{M}) - h_1(\mathbb{M}) + 1$ if and only if depth $gr_{\mathbb{M}}(M) = 0$.

  (b) $e_2(\mathbb{M}) = e_0(\mathbb{M}) - h_0(\mathbb{M}) - h_1(\mathbb{M}) + 2$ if and only if depth $gr_{\mathbb{M}}(M) \geq 1$.
  Further, in case (a), $gr_{\widetilde{\mathbb{M}}}(M)$ is Cohen–Macaulay; in case (b),

$$P_{\mathbb{M}}(z) = \frac{h_0(\mathbb{M}) + h_1(\mathbb{M})z + h_2(\mathbb{M})z^2 + z^3}{(1-z)^2}.$$

Proof. We have $e_i(\mathbb{M}) = e_i(\widetilde{\mathbb{M}})$ for $i = 0,1,2$

$$h_0(\widetilde{\mathbb{M}}) = \lambda(M/\widetilde{M}_1) = \lambda(M/M_1) = h_0(M)$$

and

$$\begin{aligned}
h(\widetilde{\mathbb{M}}) &= \lambda(\widetilde{M}_1/JM + \widetilde{M}_2) = \lambda(M_1/JM + \widetilde{M}_2) \\
&= \lambda(M_1/JM + M_2) - \lambda(JM + \widetilde{M}_2/JM + M_2) \\
&= h(\mathbb{M}) - \lambda\left(\widetilde{M}_2/M_2 + (\widetilde{M}_2 \cap JM)\right) \\
&= h(\mathbb{M}) - \lambda(\widetilde{M}_2/M_2).
\end{aligned}$$

Since $M_2 \cap JM \subseteq \widetilde{M}_2 \cap JM = JM_1$, we also have $M_2 \cap JM = JM_1$, which implies by (2.15)

$$h(\mathbb{M}) = h_1(\mathbb{M}).$$

Further

$$2e_0(\mathbb{M}) - 2h_0(\mathbb{M}) - h(\mathbb{M}) + 1 = e_1(\mathbb{M}) = e_1(\widetilde{\mathbb{M}})$$
$$\geq 2e_0(\widetilde{\mathbb{M}}) - 2h_0(\widetilde{\mathbb{M}}) - h(\widetilde{\mathbb{M}})$$
$$= 2e_0(\mathbb{M}) - 2h_0(\mathbb{M}) - h(\mathbb{M}) + \lambda(\widetilde{M}_2/M_2)$$

so that $0 \leq \lambda(\widetilde{M}_2/M_2) \leq 1$.

If $\widetilde{M}_2 \neq M_2$, then $\lambda(\widetilde{M}_2/M_2) = 1$ and

$$e_1(\widetilde{\mathbb{M}}) = 2e_0(\widetilde{\mathbb{M}}) - 2h_0(\widetilde{\mathbb{M}}) - h(\widetilde{\mathbb{M}})$$

so that $gr_{\widetilde{\mathbb{M}}}(M)$ is Cohen–Macaulay by Theorem 2.8. This proves (1).

Let us prove (2). We have $r = 2$ and depth $gr_{\widetilde{\mathbb{M}}}(M) \geq 1 = r - 1$ so that, by Theorem 2.5,

$$e_1(\mathbb{M}) = e_1(\widetilde{\mathbb{M}}) = \sum_{j \geq 0} v_j(\widetilde{\mathbb{M}}), \quad e_2(\mathbb{M}) = e_2(\widetilde{\mathbb{M}}) = \sum_{j \geq 1} j v_j(\widetilde{\mathbb{M}}).$$

Now

$$v_0(\widetilde{\mathbb{M}}) = \lambda(\widetilde{M}_1/JM) = \lambda(M_1/JM) = e_0(\mathbb{M}) - h_0(\mathbb{M})$$
$$v_1(\widetilde{\mathbb{M}}) = \lambda(\widetilde{M}_2/J\widetilde{M}_1) = \lambda(\widetilde{M}_2/JM_1) = \lambda(\widetilde{M}_2/M_2) + \lambda(M_2/JM_1)$$
$$= \lambda(\widetilde{M}_2/M_2) + e_0(\mathbb{M}) - h_0(\mathbb{M}) - h_1(\mathbb{M})$$
$$= \lambda(\widetilde{M}_2/M_2) + e_0(\mathbb{M}) - h_0(\mathbb{M}) - h(\mathbb{M}),$$

where we used the equality $e_0(\mathbb{M}) = h_0(\mathbb{M}) + h_1(\mathbb{M}) + \lambda(M_2/JM_1)$ proved in Proposition 2.1.

This implies

$$2e_0(\mathbb{M}) - 2h_0(\mathbb{M}) - h(\mathbb{M}) + 1 = e_1(\mathbb{M}) = e_1(\widetilde{\mathbb{M}})$$
$$= v_0(\widetilde{\mathbb{M}}) + v_1(\widetilde{\mathbb{M}}) + \sum_{j \geq 2} v_j(\widetilde{\mathbb{M}})$$
$$= \lambda(\widetilde{M}_2/M_2) + 2e_0(\mathbb{M}) - 2h_0(\mathbb{M}) - h(\mathbb{M}) + \sum_{j \geq 2} v_j(\widetilde{\mathbb{M}})$$

so that

$$\lambda(\widetilde{M}_2/M_2) + \sum_{j \geq 2} v_j(\widetilde{\mathbb{M}}) = 1.$$

In the case $\sum_{j \geq 2} v_j(\widetilde{\mathbb{M}}) = 1$, we have $\widetilde{M}_2 = M_2$ and $e_1(\mathbb{M}) = v_0(\widetilde{\mathbb{M}}) + v_1(\widetilde{\mathbb{M}}) + 1$. We claim that this implies $M_3 \neq JM_2$ and $v_2(\widetilde{\mathbb{M}}) = 1$. Namely, if $M_3 = JM_2$, then $\mathfrak{q}^2 L = J\mathfrak{q}L$ so that $M_{j+1} = JM_j$ for every $j \geq 2$. Since $M_2 \cap JM = JM_1$, by

Valabrega–Valla $gr_{\mathbb{M}}(M)$ is Cohen–Macaulay with $v_j(\mathbb{M}) = 0$ for every $j \geq 2$. But then $e_1(\mathbb{M}) = v_0(\mathbb{M}) + v_1(\mathbb{M})$, a contradiction.

Hence $M_3 \neq JM_2$, so that

$$J\widetilde{M_2} = JM_2 \subset M_3 \subseteq \widetilde{M_3}$$

and $v_2(\widetilde{\mathbb{M}}) = 1$. This proves the claim. Now we can write

$$e_2(\mathbb{M}) = e_2(\widetilde{\mathbb{M}}) = v_1(\widetilde{\mathbb{M}}) + \sum_{j \geq 2} j v_j(\widetilde{\mathbb{M}})$$

$$= \lambda(\widetilde{M_2}/M_2) + e_0(\mathbb{M}) - h_0(\mathbb{M}) - h_1(\mathbb{M}) + \sum_{j \geq 2} j v_j(\widetilde{\mathbb{M}})$$

$$= e_0(\mathbb{M}) - h_0(\mathbb{M}) - h_1(\mathbb{M}) + 1 + \sum_{j \geq 2}(j-1)v_j(\widetilde{\mathbb{M}}).$$

Hence we have only two possibilities for $e_2(\mathbb{M})$, namely

$$e_2(\mathbb{M}) = \begin{cases} e_0(\mathbb{M}) - h_0(\mathbb{M}) - h_1(\mathbb{M}) + 1 & \text{if } \sum_{j \geq 2} v_j(\widetilde{\mathbb{M}}) = 0, \\ e_0(\mathbb{M}) - h_0(\mathbb{M}) - h_1(\mathbb{M}) + 2 & \text{otherwise.} \end{cases} \quad (3.3)$$

Now, if depth $gr_{\mathbb{M}}(M) \geq 1$, then $\widetilde{M_2} = M_2$, hence $\sum_{j \geq 2} v_j(\widetilde{\mathbb{M}}) = 1$ and we have $e_2(\mathbb{M}) = e_0(\mathbb{M}) - h_0(\mathbb{M}) - h_1(\mathbb{M}) + 2$.

Conversely, if $e_2(\mathbb{M}) = e_0(\mathbb{M}) - h_0(\mathbb{M}) - h_1(\mathbb{M}) + 2$, then $v_2(\widetilde{\mathbb{M}}) = 1$ and $v_j(\widetilde{\mathbb{M}}) = 0$ for every $j \geq 3$. This implies

$$1 = \lambda(\widetilde{M_3}/J\widetilde{M_2}) = \lambda(\widetilde{M_3}/JM_2) \geq \lambda(M_3/JM_2) \geq 1,$$

so that $\widetilde{M_3} = M_3$. Hence, since $v_j(\widetilde{\mathbb{M}}) = 0$ for every $j \geq 3$, we get $\widetilde{M_j} = M_j$ for every $j \geq 0$, which is equivalent to depth $gr_{\mathbb{M}}(M) \geq 1$.

This proves (a); as for (b), it follows by (a) and (3.3). We come now to the last assertions of the theorem.

In case (a), we have $\widetilde{M_2} \neq M_2$, so that $gr_{\widetilde{\mathbb{M}}}(M)$ is Cohen–Macaulay by (1). In case (b), $\mathbb{M} = \widetilde{\mathbb{M}}$ so that $v_2(\mathbb{M}) = v_2(\widetilde{\mathbb{M}}) = 1$ and $v_j(\mathbb{M}) = v_j(\widetilde{\mathbb{M}}) = 0$ for every $j \geq 3$. Since by Theorem 2.5 we have $e_i(\mathbb{M}) = \sum_{j \geq i-1} \binom{j}{i-1} v_j(\mathbb{M})$, we get $e_3(\mathbb{M}) = 1$ and $e_j(\mathbb{M}) = 0$ for every $j \geq 4$. These values of the e_i's give the required Hilbert series. $\qquad \square$

The assumptions $\widetilde{M_1} = M_1$ and $\widetilde{M_2} \cap JM = JM_1$ in the above theorem seem very strong, but they are satisfied by the q-adic filtration of any primary integrally closed ideal q, in particular by the m-adic filtration.

Corollary 3.2. *Let* q *be an* m-*primary ideal in the Cohen–Macaulay local ring* A *of dimension* r *and* $\mathbb{M}$ *the* q-*adic filtration on* A. *If* q *is integrally closed and*

$e_1(\mathbb{M}) = 2e_0(\mathbb{M}) - 2h_0(\mathbb{M}) - h(\mathbb{M}) + 1$, *the following conditions are equivalent and each implies*

$$P_\mathbb{M}(z) = \frac{h_0(\mathbb{M}) + h_1(\mathbb{M})z + h_2(\mathbb{M})z^2 + z^3}{(1-z)^r}.$$

(a) depth $gr_\mathbb{M}(M) \geq r - 1$.
(b) $e_2(\mathbb{M}) = e_0(\mathbb{M}) - h_0(\mathbb{M}) - h_1(\mathbb{M}) + 2$.

If this is not the case, then $e_2(\mathbb{M}) = e_0(\mathbb{M}) - h_0(\mathbb{M}) - h_1(\mathbb{M}) + 1$.

Proof. Since $\mathfrak{q} \subseteq \widetilde{\mathfrak{q}} \subseteq \overline{\mathfrak{q}}$, we have $\mathfrak{q} = \widetilde{\mathfrak{q}}$; on the other hand, if J is an ideal generated by a maximal sequence of $\mathbb{M}$-superficial elements for $\mathfrak{q}$, by a result of Huneke and Itoh (see [54] and [58]), we have

$$\widetilde{\mathfrak{q}^2} \cap J \subseteq \overline{\mathfrak{q}^2} \cap J = J\overline{\mathfrak{q}} = J\mathfrak{q}$$

so that

$$\widetilde{\mathfrak{q}^2} \cap J = J\mathfrak{q}.$$

Hence the equivalence between (a) and (b) follows by the theorem if $r = 2$. When $r \geq 3$, by a result of Itoh (see [57]), we can find an ideal $\mathfrak{a}$ generated by an $\mathbb{M}$-superficial sequence for $\mathfrak{q}$ of length $r - 2$ such that $\mathfrak{q}/\mathfrak{a}$ is integrally closed. Then we have $e_i(\mathbb{M}) = e_i(\mathbb{M}/\mathfrak{a})$ for $i = 0, 1, 2$,

$$h_0(\mathbb{M}) = \lambda(A/J) = h_0(\mathbb{M}/\mathfrak{a})$$

and

$$h_1(\mathbb{M}) = h(\mathbb{M}) = \lambda(A/J + \mathfrak{q}^2) = h(\mathbb{M}/\mathfrak{a}) = h_1(\mathbb{M}/\mathfrak{a}).$$

Hence all the assumptions of the theorem hold for the two-dimensional Cohen–Macaulay local ring $A/\mathfrak{a}$ and the integrally closed primary ideal $\mathfrak{q}/\mathfrak{a}$. The equivalence between (a) and (b) follows by the theorem because, by Sally's machine, depth $gr_\mathbb{M}(A) \geq r - 1$ if and only if depth $gr_\mathbb{M}(A/\mathfrak{a}) \geq 1$.

As for the last assertion if depth $gr_\mathbb{M}(M) < r - 1$, then by using Sally's machine, we deduce depth $gr_\mathbb{M}(A/\mathfrak{a}) = 0$. Since $\mathfrak{q}/\mathfrak{a}$ is integrally closed and $A/\mathfrak{a}$ is a two-dimensional local Cohen–Macaulay ring, we may apply Theorem 3.4 and it is easy to see that $e_2(\mathbb{M}) = e_0(\mathbb{M}) - h_0(\mathbb{M}) - h_1(\mathbb{M}) + 1$ since the integers involved do not change passing to $\mathbb{M}/\mathfrak{a}$. $\qquad\square$

Chapter 4
Sally's Conjecture and Applications

Let $\mathfrak{q}$ be an $\mathfrak{m}$-primary ideal of A and let M be a Cohen–Macaulay A-module of dimension r. Consider the good $\mathfrak{q}$-filtration $\mathbb{M} = \mathbb{M}_L$ induced by a submodule L of M (see (2.18)). In Theorem 2.9 we proved that, if $\mathbb{M}$ has minimal multiplicity, namely $e_0(\mathbb{M}) = h_0(\mathbb{M}) + h_1(\mathbb{M})$, then

$$P_{\mathbb{M}}(z) = \frac{h_0(\mathbb{M}) + h_1(\mathbb{M})z}{(1-z)^r}$$

and $gr_{\mathbb{M}}(M)$ is Cohen–Macaulay.

In the "next case", when $e_0(\mathbb{M}) = h_0(\mathbb{M}) + h_1(\mathbb{M}) + 1$, we say that $\mathbb{M}$ has almost minimal multiplicity. Almost minimal multiplicity is much more difficult to handle, even for the $\mathfrak{m}$-adic filtration on a Cohen–Macaulay local ring. In this particular case $h_0(\mathfrak{m}) = 1$ and $h_1(\mathfrak{m}) = \mu(\mathfrak{m}) - r = h$, the embedding codimension. Hence, in this case, almost minimal multiplicity means

$$e_0(\mathfrak{m}) = h + 2$$

For example the Cohen–Macaulay one-dimensional local ring $A = k[[t^4, t^5, t^{11}]]$ has almost minimal multiplicity (with respect the $\mathfrak{m}$-adic filtration) and its Hilbert series is

$$P_{\mathfrak{m}}(z) = \frac{1 + hz + z^3}{(1-z)},$$

but the associated graded ring is not Cohen–Macaulay.

It was conjectured by Sally in [91] that, for an r-dimensional Cohen–Macaulay local ring, and with respect to the $\mathfrak{m}$-adic filtration, almost minimal multiplicity forces the depth of the associated graded ring to be at least $r - 1$. After 13 years, the conjecture was proved by Wang in [115] and, at the same time, by Rossi and Valla in [80]. In particular it was proved that an r-dimensional Cohen–Macaulay local ring A has almost minimal multiplicity if and only if for some integer s such that $2 \le s \le e_0 - 1$

$$P_A(z) = \frac{1 + hz + z^s}{(1-z)^r}$$

M.E. Rossi and G. Valla, *Hilbert Functions of Filtered Modules*, Lecture Notes of the Unione Matematica Italiana 9, DOI 10.1007/978-3-642-14240-6_4, © Springer-Verlag Berlin Heidelberg 2010

Later the conjecture was stated for any m-primary ideal of a Cohen–Macaulay ring and an extended version was proved in [16], [22], [50] and [76] by following Rossi and Valla's proof.

In this chapter we present a proof of this result in the general case of a module endowed with the filtration induced by L. The crucial point of this result is a bound on the reduction number of $\mathbb{M}$.

As we have already seen in Chap. 1, if J is an ideal generated by a maximal $\mathbb{M}$-superficial sequence for q, then $\mathbb{M}$ is a good J-filtration and hence for large n we have

$$M_{n+1} = JM_n.$$

In particular J is a minimal $\mathbb{M}$-reduction of q. Following the classical theory of reductions of an ideal, we denote by

$$r_J(\mathbb{M}) := \min\{n \in \mathbf{N} \mid M_{j+1} = JM_j \text{ for every } j \geq n\}$$

the reduction number of $\mathbb{M}$ with respect to J. Since $\mathbb{M} = \mathbb{M}_L$ for a given submodule L of M, we clearly have

$$r_J(\mathbb{M}) := \min\{n \in \mathbf{N} \mid M_{n+1} = JM_n\} = \min\{n \in \mathbf{N} \mid v_n(\mathbb{M}) = 0\}.$$

If $\mathbb{M}$ is the q-adic filtration on the ring A, then we write $r_J(q)$ instead of $r_J(\mathbb{M})$. In the one-dimensional case

$$r_J(\mathbb{M}) \leq e_0(\mathbb{M}) - 1.$$

This bound can be easily extended to higher dimensions under the assumption depth $gr_{\mathbb{M}}(M) \geq r - 1$. Moreover, as in the classical case (see [101] and [90]), if M is Cohen–Macaulay and depth $gr_{\mathbb{M}}(M) \geq r - 1$, then $r_J(\mathbb{M})$ is independent of J. In Theorem 4.3 we will prove that if $r = \dim M = 2$ or, more in general, if we assume depth $gr_{\mathbb{M}}(M) \geq r - 2$, then

$$r_J(\mathbb{M}) \leq e_1(\mathbb{M}) - e_0(\mathbb{M}) + h_0(\mathbb{M}) + 1.$$

By using this bound, as a bonus, we get easy proofs of new "border cases" theorems. In particular we consider the filtrations having $e_1(\mathbb{M}) = e_0(\mathbb{M}) - h_0(\mathbb{M}) + 1$ or $e_1(\mathbb{M}) = e_0(\mathbb{M}) - h_0(\mathbb{M}) + 2$. If $e_1(\mathbb{M}) = e_0(\mathbb{M}) - h_0(\mathbb{M}) + 1$, we have

$$P_{\mathbb{M}}(z) = \frac{h_0(\mathbb{M}) + h_1(\mathbb{M})z + z^2}{(1 - z)^r}$$

if $M_2 \cap JM = JM_1$ (Theorem 2.10) or $e_2(\mathbb{M}) \neq 0$ (Theorem 3.3). In this case $\mathbb{M}$ has almost minimal multiplicity. The case when $M_2 \cap JM \neq JM_1$ and $e_2(\mathbb{M}) = 0$ is more difficult and, by using the usual approach, we reprove in dimension two a nice result due to Goto, Nishida and Ozeki (see [33]).

4.1 A Bound on the Reduction Number

Let q be an $\mathfrak{m}$-primary ideal of A and let $\mathbb{M} = \{M_j\}_{j \geq 0}$ be a good q-filtration of a finitely generated A-module M. We will denote by $\widetilde{\mathbb{M}} = \{\widetilde{M}_j\}_{j \geq 0}$ the Ratliff–Rush filtration (see Chap. 3). Since $\widetilde{\mathbb{M}}$ is a good q-filtration and $M_j \subseteq \widetilde{M}_j$, we remark that $\oplus_{j \geq 0}(\widetilde{M}_j/M_j)$ has a canonical structure as a graded module over the Rees algebra $\mathscr{R}(q) = A[qT] = \oplus_{j \geq 0} q^j T^j$.

Now we prove a result on the reduction number of $\mathbb{M}$ extending to modules an inequality which was proved in [75, Theorem 1.3], in the case of the q-adic filtration on A.

As usual we denote by J an ideal generated by a maximal $\mathbb{M}$-superficial sequence for q. Then

$$N := \oplus_{j \geq 0}(\widetilde{M}_{j+1}/J\widetilde{M}_j + M_{j+1})$$

has a canonical structure as a graded $\mathscr{R}(J)$-module. Assuming $\operatorname{depth}_q M \geq 1$, by Lemma 3.1, there exists an integer p such that $M_j = \widetilde{M}_j$ for every $j \geq p$. This implies that

$$\widetilde{M}_{j+1} = J\widetilde{M}_j + M_{j+1}$$

for every $j \geq p$. Hence N is an A-module of finite length. In the following we denote by

$$v := \dim_k N/\mathfrak{m}N$$

the minimal number of generators of N as A-module. Next result controls the reduction number of the q-adic filtration by using any good q-filtration of an A-module M.

Theorem 4.1. *Let $\mathbb{M}$ be a good q-filtration on the A-module M of positive depth and let J be an ideal generated by a maximal $\mathbb{M}$-superficial sequence for q. Then*

$$q^v \subseteq Jq^{v-1} + (M_{v+n} :_A \widetilde{M}_n)$$

for every positive integer n.

Proof. Let p be an integer such that $\widetilde{M}_n = J\widetilde{M}_{n-1} + M_n$ for all $n > p$. For all $n = 1, \ldots, p$ we consider the elements $m_{1n}, \ldots, m_{v_n n} \in \widetilde{M}_n$ such that the corresponding elements in $N_n = \widetilde{M}_n/J\widetilde{M}_{n-1} + M_n$ form a minimal system of generators as an A-module. In particular if we define the submodules

$$L_n := <m_{1n}, \ldots, m_{v_n n}> \subseteq \widetilde{M}_n,$$

then $\widetilde{M}_n = J\widetilde{M}_{n-1} + L_n + M_n$ (with $L_n = 0$ if $n > p$). It is easy to see that for every $n \geq 1$ we can write

$$\widetilde{M}_n = \sum_{j=0}^{n} J^{n-j}L_j + M_n.$$

We have $v = \sum_{n=1}^{p} v_n$ and so $|(in)| = v$ if $i = 1, \ldots, v_n$ and $n = 1, \ldots, p$.

Denote by a_{in} an element of $\mathfrak{q}$, then

$$(a_{in}T)m_{in} \in \widetilde{M}_{n+1} = \sum_{j=0}^{n+1} J^{n+1-j}L_j + M_{n+1}.$$

Then there exist $c_{(in)(kj)} \in J^{n+1-j}$ and $\alpha_{in+1} \in M_{n+1}$ such that

$$(a_{in}T)m_{in} = \sum_{j=1}^{n+1} \sum_{k=1}^{v_j} c_{(in)(kj)} T^{n+1-j} m_{kj} + \alpha_{in+1}$$

with $m_{k\,p+1} = 0$ for every k.

Thus if we consider the relations in $\oplus_{j \geq 0}(\widetilde{M}_j/M_j)$

$$\sum_{j=1}^{n+1} \sum_{k=1}^{v_j} c_{(in)(kj)} T^{n+1-j}\overline{m_{kj}} - (a_{in}T)\overline{m_{in}} = \overline{0}$$

we get a system of v linear equations in the v variables $\overline{m_{kj}}$ where $j = 1, \ldots, p$ and $k = 1, \ldots, v_j$.

The corresponding matrix B of the coefficients of the variables has size $v \times v$ and entries which are homogeneous elements in the Rees ring $\mathscr{R}(\mathfrak{q})$. Since the $(in)(kj)$-entry has degree $n+1-j$ if $n+1 \geq j$ and is zero otherwise, we may assign degree $n+1-j$ to the $(in)(kj)$-entry of B whatsoever. This implies that every two by two minor of B is an homogenous element, hence its determinant $\det(B)$ is homogeneous too and its degree is v because the elements on the diagonal $(in) = (kj)$, which are $(c_{(in)(in)} - a_{in})T$, all have degree 1.

If $a = \prod a_{in}$ for $n = 0, \ldots, p$ and $i = 1, \ldots, v_n$, it is easy to see that

$$\det(B) = (-1)^{v}(a - \sigma)T^{v}$$

for a suitable $\sigma \in J\mathfrak{q}^{v-1}$.

Now the Cayley–Hamilton theorem asserts that $\det(B)$ kills all the variables. Hence for all $i = 1, \ldots, v_n$ and $n \geq 1$

$$\det(B)\overline{m_{in}} = \overline{0} \in \widetilde{M}_{v+n}/M_{v+n},$$

so that

$$a \in J\mathfrak{q}^{v-1} + (M_{v+n} :_A \widetilde{M}_n)$$

for every $n \geq 1$. We may repeat the same procedure for all monomial $a = \prod a_{in}$ in $\mathfrak{q}^{v}$ and the result follows. $\qquad\qquad\square$

Since $M_{n+1} = JM_n = J\widetilde{M}_n$ for large n, we may define the integer

$$k := \min\{t \mid M_{t+1} \subseteq J\widetilde{M}_t\}.$$

Corollary 4.1. *Let* $\mathbb{M}$ *be a good* $\mathfrak{q}$-*filtration on the A-module M of positive depth and let J be an ideal generated by a maximal* $\mathbb{M}$-*superficial sequence for* $\mathfrak{q}$. *Then*

$$\mathfrak{q}^v M_{k+1} \subseteq J M_{k+v}.$$

Proof. From the above theorem we get

$$\mathfrak{q}^v M_{k+1} \subseteq (J\mathfrak{q}^{v-1} + M_{v+k} :_A \widetilde{M}_k) M_{k+1} \subseteq J M_{k+v} + (M_{v+k} :_A \widetilde{M}_k) J\widetilde{M}_k \subseteq J M_{k+v}.$$

$\square$

Corollary 4.2. *Let L be a submodule of a module M of positive depth and let* $\mathbb{M} = \mathbb{M}_L$ *be the good* $\mathfrak{q}$-*filtration on M induced by L. Let J be an ideal generated by a maximal* $\mathbb{M}$-*superficial sequence for* $\mathfrak{q}$. *With the above notation we have*

$$r_J(\mathbb{M}) \le k + v.$$

Proof. We remark that, by the definition of $\mathbb{M}_L$, we have $M_{k+v+1} = \mathfrak{q}^v M_{k+1}$. Then the result follows by Corollary 4.1. $\square$

In the following we denote by

$$S_J := \{n \in \mathbb{N} \mid M_{j+1} \cap J\dot{\overset{..}{M}}_j = JM_j \text{ for every } j, \ 0 \le j \le n\}.$$

We remark that $S_J \ne \emptyset$ since $0 \in S_J$.

The following result extends Theorem 1.3. in [75] to modules. We include here a proof even if it is essentially a natural recasting of the original result proved in the classical case.

Theorem 4.2. *Let L be a submodule of the module M of positive depth and let* $\mathbb{M} = \mathbb{M}_L$ *be the good* $\mathfrak{q}$-*filtration on M induced by L. Let J be an ideal generated by a maximal* $\mathbb{M}$-*superficial sequence for* $\mathfrak{q}$. *If* $n \in S_J$, *then*

$$r_J(\mathbb{M}) \le \sum_{i \ge 0} v_i(\widetilde{\mathbb{M}}) + n + 1 - \sum_{i=0}^{n} v_i(\mathbb{M})$$

Proof. By Corollary 4.2 we have

$$r_J(\mathbb{M}) \le v + k = \sum_{i \ge 0} v_i + k \le \sum_{i \ge 0} \lambda(\widetilde{M}_{i+1}/J\widetilde{M}_i + M_{i+1}) + k.$$

But it is clear that

$$\lambda(\widetilde{M}_{i+1}/J\widetilde{M}_i + M_{i+1}) = v_i(\widetilde{\mathbb{M}}) - \lambda(M_{i+1}/J\widetilde{M}_i \cap M_{i+1}) \leq v_i(\widetilde{\mathbb{M}}),$$

so that

$$\lambda(\widetilde{M}_{i+1}/J\widetilde{M}_i + M_{i+1}) = v_i(\widetilde{\mathbb{M}}) - v_i(\mathbb{M}) \quad \text{if } 0 \leq i \leq n$$

and $\lambda(\widetilde{M}_{i+1}/J\widetilde{M}_i + M_{i+1}) < v_i(\widetilde{\mathbb{M}})$ if $0 \leq i \leq k-1$. This implies

$$r_J(\mathbb{M}) \leq \sum_{i=0}^{n}(v_i(\widetilde{\mathbb{M}}) - v_i(\mathbb{M})) + \sum_{i \geq n+1} v_i(\widetilde{\mathbb{M}}) + k$$

so that the conclusion follows if $k \leq n+1$. If $k \geq n+2$, then we have

$$r_J(I) \leq \sum_{i=0}^{n}(v_i(\widetilde{\mathbb{M}}) - v_i(\mathbb{M})) + \sum_{i=n+1}^{k-1}(v_i(\widetilde{\mathbb{M}}) - 1) + \sum_{i \geq k} v_i(\widetilde{\mathbb{M}}) + k$$

$$= \sum_{i \geq 0} v_i(\widetilde{\mathbb{M}}) + n + 1 - \sum_{i=0}^{n} v_i(\mathbb{M}).$$

$\square$

A nice application of the above theorem is given by Kinoshita, Nishida, Sakata, Shinya in [59].

In some cases we can get rid of the v_i's in the formula given in Theorem 4.2.

Theorem 4.3. *Let L be a submodule of the r-dimensional Cohen–Macaulay module M and let $\mathbb{M} = \mathbb{M}_L$ be the good $\mathfrak{q}$-filtration on M induced by L. If depth $gr_{\mathbb{M}}(M) \geq r-2$, then*

$$r_J(\mathbb{M}) \leq e_1(\mathbb{M}) - e_0(\mathbb{M}) + h_0(\mathbb{M}) + 1$$

for every ideal J generated by a maximal $\mathbb{M}$-superficial sequence for $\mathfrak{q}$.

Proof. First of all we prove that we may reduce the problem to dimension $r \leq 2$. Let $r > 2$ and $J = (a_1, \ldots, a_r)$ be an ideal generated by a maximal $\mathbb{M}$-superficial sequence for $\mathfrak{q}$; we consider the ideal $K := (a_1, \ldots, a_{r-2})$. Since depth $gr_{\mathbb{M}}(M) \geq r-2$, by Lemma 1.3 and the Valabrega–Valla criterion, we get $M_{n+1} \cap KM = KM_n$ for every n, which easily implies

$$v_j(\mathbb{M}) = v_j(\mathbb{M}/KM)$$

for every j. Hence

$$r_J(\mathbb{M}) = r_{\mathfrak{a}}(\mathbb{M}/KM)$$

where $\mathfrak{a} = (a_{r-1}, a_r)$. Moreover, $e_1(\mathbb{M}) = e_1(\mathbb{M}/KM)$, $e_0(\mathbb{M}) = e_0(\mathbb{M}/KM)$ and $h_0(\mathbb{M}) = h_0(\mathbb{M}/KM)$ since $KM \subseteq L$ by definition.

Hence we may assume $r \leq 2$; by Theorem 4.2 with $n = 0$, we have

$$r_J(\mathbb{M}) \leq \sum_{i \geq 0} v_i(\widetilde{\mathbb{M}}) + 1 - v_0(\mathbb{M}) = \sum_{i \geq 0} v_i(\widetilde{\mathbb{M}}) + 1 - e_0(\mathbb{M}) + h_0(\mathbb{M}).$$

Since depth $gr_{\widetilde{\mathbb{M}}}(M) \geq 1 \geq r - 1$, by Theorem 2.5 we have $\sum_{i \geq 0} v_i(\widetilde{\mathbb{M}}) = e_1(\widetilde{\mathbb{M}}) = e_1(\mathbb{M})$. The conclusion follows.   □

Remark 4.1. We note that the above bound is sharp and gives easy proofs of some results presented in Chap. 2. For example if we consider $e_1(\mathbb{M}) = e_0(\mathbb{M}) - h_0(\mathbb{M})$, i.e. the minimum value of $e_1(\mathbb{M})$ with respect to Northcott's bound, we immediately get $r_J(\mathbb{M}) \leq 1$, hence $M_2 = JM_1$ and obviously $gr_{\mathbb{M}}(M)$ is Cohen–Macaulay (see Theorem 2.9).

If $e_1(\mathbb{M}) = e_0(\mathbb{M}) - h_0(\mathbb{M}) + 1$, then $r_J(\mathbb{M}) \leq 2$, hence $M_3 = JM_2$ and, if $M_2 \cap J = JM_1$, by the Valabrega–Valla criterion, we conclude that $gr_{\mathbb{M}}(M)$ is Cohen–Macaulay (see Theorem 2.10).

4.2 A Generalization of Sally's Conjecture

As a consequence of Theorem 4.3, we present now an extended version of the Sally conjecture. We recall that in the case of the $\mathfrak{m}$-adic filtration, the question raised by Sally was the following: if A is Cohen–Macaulay of dimension r and it has almost minimal multiplicity, that is $e_0(\mathfrak{m}) = \mu(\mathfrak{m}) - r + 2$, is it true that depth $gr_{\mathfrak{m}}(A) \geq r - 1$?

Sally's conjecture was proved in [80] and independently in [115]. The next theorem is a generalization of this result.

Let $\mathbb{M}$ be a good q-filtration of M and J an ideal generated by a maximal sequence of $\mathbb{M}$-superficial elements for q. For every integer $j \geq 0$, we have defined the integers

$$v_j(\mathbb{M}) = \lambda(M_{j+1}/JM_j)$$
$$vv_j(\mathbb{M}) = \lambda(M_{j+1} \cap JM/JM_j)$$
$$w_j(\mathbb{M}) = \lambda(M_{j+1}/M_{j+1} \cap JM).$$

so that we have the formula

$$v_j(\mathbb{M}) = w_j(\mathbb{M}) + vv_j(\mathbb{M}).$$

Further, if $x \in J$ is a superficial element, it is easy to see that

$$vv_j(\mathbb{M}) \geq vv_j(\mathbb{M}/xM)$$
$$w_j(\mathbb{M}) = w_j(\mathbb{M}/xM)$$
$$v_j(\mathbb{M}) \geq v_j(\mathbb{M}/xM)$$

for every $j \geq 0$.

We recall that if $vv_j(\mathbb{M}) = 0$ for every $j \geq 0$, then by the Valabrega–Valla criterion, $gr_{\mathbb{M}}(M)$ is Cohen–Macaulay.

Theorem 4.4. *Let L be a submodule of the r-dimensional Cohen–Macaulay module M and let $\mathbb{M} = \mathbb{M}_L$ be the good q-filtration on M induced by L. Let J be an ideal generated by a maximal $\mathbb{M}$-superficial sequence for* q. *If there exists a positive integer p such that:*

(1) $vv_j(\mathbb{M}) = 0$ for every $j \leq p - 1$
and
(2) $v_p(\mathbb{M}) \leq 1$,
then depth $gr_{\mathbb{M}}(M) \geq r - 1$.

Proof. Conditions (1) and (2) are preserved modulo superficial elements in J, so that, by using the Sally machine, we may reduce the problem to the case $r = 2$ and we have to prove that depth $gr_{\mathbb{M}}(M) \geq 1$.

We may assume that $v_p(\mathbb{M}) = \lambda(M_{p+1}/JM_p) = 1$, otherwise, by the Valabrega–Valla criterion, we immediately get that $gr_{\mathbb{M}}(M)$ is Cohen–Macaulay.

Since $M_{p+1} = qM_p$ is generated over A by the products am with $a \in q$ and $m \in M_p$, the condition $\lambda(M_{p+1}/JM_p) = 1$ implies $M_{p+1} = JM_p + (a)m$ with $a \in q$, $m \in M_p$ and $am \notin JM_p$. Then for every $n \geq p$ the multiplication by a gives a surjective map from M_{n+1}/JM_n to M_{n+2}/JM_{n+1}; this implies

$$v_n(\mathbb{M}) = \lambda(M_{n+1}/JM_n) \leq 1$$

for every $n \geq p$.

Let $J = (x, y)$ and $s = r_{(y)}(\mathbb{M}/xM)$ be the reduction number of $\mathbb{M}/xM$ with respect to the ideal (y). This means that

$$v_j(\mathbb{M}/xM) = 0 \quad \text{if} \quad j \geq s, \quad v_j(\mathbb{M}/xM) > 0 \quad \text{if} \quad j < s.$$

It follows that, when $s \leq p$, we get $vv_j(\mathbb{M}/xM) = 0$ for every $j \geq 0$, so that $gr_{\mathbb{M}/xM}(M/xM)$ is Cohen–Macaulay. By Sally's machine, this implies $gr_{\mathbb{M}}(M)$ is Cohen–Macaulay as well.

Hence we may assume $s > p \geq 1$ and we prove that

$$v_j(\mathbb{M}) = v_j(\mathbb{M}/xM) \tag{4.1}$$

for $j = 0, \ldots, s - 1$. If $0 \leq j \leq p - 1$, we have $vv_j(\mathbb{M}) = 0$ by assumption, so that

$$v_j(\mathbb{M}) = w_j(\mathbb{M}) = w_j(\mathbb{M}/xM) \leq v_j(\mathbb{M}/xM) \leq v_j(\mathbb{M}).$$

On the other hand, if $p \leq j \leq s - 1$, we have

$$0 < v_j(\mathbb{M}/xM) \leq v_j(\mathbb{M}) \leq 1.$$

This proves (4.1) and also that

$$v_j(\mathbb{M}) = v_j(\mathbb{M}/xM) = 1$$

for all $p \leq j \leq s-1$.

Now, for every $j \geq 0$, we have $v_j(\mathbb{M}/xM) = \lambda(M_{j+1}/JM_j + x(M_{j+1}:x))$, and hence for $j = 0, \ldots, s-1$

$$v_j(\mathbb{M}) = v_j(\mathbb{M}/xM) = \lambda(M_{j+1}/JM_j + x(M_{j+1}:x)).$$

From the above equality and the following exact sequence

$$0 \longrightarrow M_j : x/M_j : J \overset{y}{\longrightarrow} M_{j+1} : x/M_j \overset{x}{\longrightarrow} M_{j+1}/JM_j \longrightarrow M_{j+1}/JM_j + x(M_{j+1}:x) \longrightarrow 0$$

we get

$$\lambda(M_j : x/M_j : J) = \lambda(M_{j+1} : x/M_j)$$

for every $j = 0, \ldots, s-1$. With $j = 1$ this gives $M_2 : x = M_1$, so that, by induction on j, we get $M_{j+1} : x = M_j$ for $j = 0, \ldots, s-1$.

We claim that if $r_J(\mathbb{M}) \leq s$, then $M_{j+1} : x = M_j$ for every $j \geq 0$ and so depth $gr_{\mathbb{M}}(M) > 0$, as required. In fact, if we have $r_J(\mathbb{M}) \leq s$, then $M_{t+1} = JM_t$ for every $t \geq s$. Let us assume by induction that $j \geq s$ and $M_j : x = M_{j-1}$ (we know that $M_s : x = M_{s-1}$), then we get

$$M_{j+1} : x = JM_j : x = (xM_j + yM_j) : x \subseteq M_j + y(M_j : x) = M_j + yM_{j-1} = M_j$$

which proves the claim.

It remains to prove that $r_J(\mathbb{M}) \leq s$. We have

$$e_1(\mathbb{M}) = e_1(\mathbb{M}/xM) = \sum_{j \geq 0} v_j(\mathbb{M}/xM) = \sum_{j \leq p-1} v_j(\mathbb{M}) + s - p.$$

Further, since $vv_j(\mathbb{M}) = 0$ for every $j \leq p-1$, we have $p-1 \in S_J$, so that, by Theorem 4.2, we get

$$r_J(\mathbb{M}) \leq \sum_{j \geq 0} v_j(\widetilde{\mathbb{M}}) + p - \sum_{j=0}^{p-1} v_j(\mathbb{M}) = e_1(\mathbb{M}) + p - \sum_{j=0}^{p-1} v_j(\mathbb{M}) = s.$$

$\square$

Corollary 4.3. *With the notation of Theorem 4.4, assume there exists a positive integer p such that:*

(1) $vv_j(\mathbb{M}) = 0$ for every $j \leq p-1$

(2) $v_p(\mathbb{M}) = 1$.

Then

$$P_{\mathbb{M}}(z) = \frac{\sum_{n=0}^{p-1} \lambda(M_n/M_{n+1} + JM_{n-1})z^n + (\lambda(M_p/JM_{p-1}) - 1)z^p + z^s}{(1-z)^r}$$

for some s, $p+1 \le s \le e_0(\mathbb{M}) - 1$. Furthermore if $M_{p+1} \cap J = JM_p$, then $gr_{\mathbb{M}}(M)$ is Cohen–Macaulay if and only if $s = p+1$.

Proof. By assumption, $M_n \cap JM = JM_{n-1}$ for every $n \le p$, then for $n < p$ we have $v_{n-1}(\mathbb{M}) - v_n(\mathbb{M}) = \lambda(M_n + JM/JM) - \lambda(M_{n+1} + JM/JM)$ $= \lambda(M_n + JM/M_{n+1} + JM) = \lambda(M_n/(M_{n+1} + JM_{n-1}))$. Further by using the information coming from the proof of the above theorem, if $p < s$, then $v_p(\mathbb{M}) = \ldots$ $= v_{s-1}(\mathbb{M}) = 1$. The Hilbert series follows by Theorem 2.5(5) and by Theorem 4.4.

It is clear that, if $gr_{\mathbb{M}}(M)$ is Cohen–Macaulay, then $s = p+1$. Conversely if $s = p+1$ and $M_{p+1} \cap J = JM_p$, we can prove that the h-polynomial of $\mathbb{M}$ coincides with that of $\mathbb{M}/JM$ and hence $gr_{\mathbb{M}}(M)$ is Cohen–Macaulay. In fact $h_n(\mathbb{M}) = h_n(\mathbb{M}/JM) = \lambda(M_n/M_{n+1} + JM_{n-1})$ for $n \le p-1$. Further $h_p(\mathbb{M}/JM)$ $= \lambda(M_p + JM/M_{p+1} + JM) = \lambda(M_p/M_{p+1} + JM_{p-1}) = \lambda(M_p/JM_{p-1})$ $- \lambda(M_{p+1}/M_{p+1} \cap J) = \lambda(M_p/JM_{p-1}) - v_p(\mathbb{M}) = h_p(\mathbb{M})$. Finally $1 = h_{p+1}(\mathbb{M})$ $= h_{p+1}(\mathbb{M}/JM)$ since $e_0(\mathbb{M}) = e_0(\mathbb{M}/JM) = \sum_{i \ge 0} h_i(\mathbb{M}/JM)$. $\qquad\square$

The assumptions of the above theorem are satisfied if we consider the m-adic filtration on a local Cohen–Macaulay ring of initial degree $p - 1$. Hence Corollary 4.3 extends Theorem 3.1 [81].

The next result is the promised *extension to modules* of the classical *Sally conjecture*. We point out that the statements (a) and (b) are independent of J. This is rare, so, when it happens is very much appreciated.

Corollary 4.4. *Let L be a submodule of the r-dimensional Cohen–Macaulay module M and let $\mathbb{M} = \mathbb{M}_L$ be the good q-filtration on M induced by L. The following conditions are equivalent:*

(a) $e_0(\mathbb{M}) = h_0(\mathbb{M}) + h_1(\mathbb{M}) + 1$

(b) $P_{\mathbb{M}}(z) = \frac{h_0(\mathbb{M}) + h_1(\mathbb{M})z + z^s}{(1-z)^r}$ for some integer $s \ge 2$.

Further, if either of the above conditions holds, then we have:

(c) depth $gr_{\mathbb{M}}(M) \ge r - 1$.

(d) Let J be the ideal generated by a maximal $\mathbb{M}$-superficial sequence and assume $M_2 \cap J = JM_1$. Then $gr_{\mathbb{M}}(M)$ is Cohen–Macaulay $\iff s = 2 \iff e_1(\mathbb{M}) = $ $= e_0(\mathbb{M}) - h_0(\mathbb{M}) + 1 \iff e_1(\mathbb{M}) = h_1(\mathbb{M}) + 2$.

Proof. It is clear that (b) implies (a). By the Abhyankar–Valla formula, if we have $e_0(\mathbb{M}) = h_0(\mathbb{M}) + h_1(\mathbb{M}) + 1$, then $\lambda(M_2/JM_1) = 1$ for every maximal superficial sequence J; hence (a) implies (b) by the above corollary, and (a) implies (c) by Theorem 4.4 with $p = 1$. Finally the first equivalence in (d) follows by Corollary 4.3 and the remaining part is a trivial computation. $\qquad\square$

We remark that the assumption $M_2 \cap J = JM_1$ in Corollary 4.4(d) is necessary. In fact if we consider Example 1.1, the q-adic filtration has almost minimal multiplicity since $\lambda(q^2/t^4q) = 1$, $P_q(z) = \frac{2+z+z^2}{(1-z)}$, but the associated graded ring $gr_q(A)$ is not Cohen–Macaulay.

By using the results of this chapter we easily recover a collection of results proved in [38, 53, 54, 58, 66] and already discussed here (see Theorems 2.9, 2.10, 4.7) by using easier methods. See also Remark 4.1 and [75, Corollary 1.9].

4.3 The Case $e_1(\mathbf{M}) = e_0(\mathbf{M}) - h_0(\mathbf{M}) + 1$

We prove now some results on $e_1(\mathbb{M})$ by using the hard machinery introduced in this chapter.

The next result completes Theorem 3.3 concerning the case $e_1(\mathbb{M}) = e_0(\mathbb{M}) - h_0(\mathbb{M}) + 1$. If we assume $\mathbb{M}$ is the m-adic filtration, in [32, Theorem 1.2], a different proof of the same result had been presented involving the structure of the Sally module.

Theorem 4.5. *Let M be a Cohen–Macaulay module of dimension two, L a submodule of M and let $\mathbb{M} = \mathbb{M}_L$ be the good q-filtration on M induced by L. Then $e_1(\mathbb{M}) = e_0(\mathbb{M}) - h_0(\mathbb{M}) + 1$ if and only if either*

$$P_{\mathbb{M}}(z) = \frac{h_0(\mathbb{M}) + h_1(\mathbb{M})z + z^2}{(1-z)^2}$$

or

$$P_{\mathbb{M}}(z) = \frac{h_0(\mathbb{M}) + h_1(\mathbb{M})z + 3z^2 - z^3}{(1-z)^2}.$$

In the first case depth $gr_{\mathbb{M}}(M) > 0$, while in the second depth $gr_{\mathbb{M}}(M) = 0$.

Proof. First we remark that if M is a Cohen–Macaulay module of dimension one and $e_1(\mathbb{M}) = e_0(\mathbb{M}) - h_0(\mathbb{M}) + 1$, then

$$P_{\mathbb{M}}(z) = \frac{h_0(\mathbb{M}) + (e_0(\mathbb{M}) - h_0(\mathbb{M}) - 1)z + z^2}{(1-z)}.$$

In fact, by Theorem 2.5, we have

$$e_1(\mathbb{M}) = \sum_{i \geq 0} v_i(\mathbb{M}) = e_0(\mathbb{M}) - h_0(\mathbb{M}) + \sum_{i \geq 1} v_i(\mathbb{M}) = e_0(\mathbb{M}) - h_0(\mathbb{M}) + 1.$$

Necessarily $v_1 = 1$ and $v_i = 0$ for every $i \geq 2$, hence the Hilbert series follows. In particular we remark that $e_2 = 1$.

Let now M be a Cohen–Macaulay module of dimension two and assume $e_1(\mathbb{M}) = e_0(\mathbb{M}) - h_0(\mathbb{M}) + 1$; if $e_2(\mathbb{M}) \neq 0$, then the result follows by Theorem 3.3.

Hence we assume $e_2(\mathbb{M}) = 0$. Since depth $gr_{\widetilde{\mathbb{M}}}(M) \geq 1$, we have $e_2(\mathbb{M}) = e_2(\widetilde{\mathbb{M}}) = \sum_{j\geq 1} j v_j(\widetilde{\mathbb{M}})$ and $e_1(\mathbb{M}) = e_1(\widetilde{\mathbb{M}}) = \sum_{j\geq 0} v_j(\widetilde{\mathbb{M}})$. It follows $v_j(\widetilde{\mathbb{M}}) = 0$ for $j \geq 1$ and $e_1(\mathbb{M}) = e_1(\widetilde{\mathbb{M}}) = v_0(\widetilde{\mathbb{M}}) = e_0(\widetilde{\mathbb{M}}) - h_0(\widetilde{\mathbb{M}})$. By Theorem 2.9 it follows that $gr_{\widetilde{\mathbb{M}}}(M)$ is Cohen–Macaulay and

$$P_{\widetilde{\mathbb{M}}}(z) = \frac{h_0(\widetilde{\mathbb{M}}) + (e_0(\mathbb{M}) - h_0(\widetilde{\mathbb{M}}))z}{(1-z)^2}. \tag{4.2}$$

Since $e_1(\mathbb{M}) = v_0(\widetilde{\mathbb{M}}) = v_0(\mathbb{M}) + \lambda(\widetilde{M}_1/M_1) = e_0(\mathbb{M}) - h_0(\mathbb{M}) + \lambda(\widetilde{M}_1/M_1)$, then $\lambda(\widetilde{M}_1/M_1) = 1$. We prove now that $\widetilde{M}_i = M_i$ for every $i \geq 2$.

First we remark that $v_1(\mathbb{M}) > 1$. In fact if $v_1(\mathbb{M}) = 1$, then $e_0(\mathbb{M}) = h_0(\mathbb{M}) + h_1(\mathbb{M}) + 1$ and by Corollary 4.4, we would have $e_2 \neq 0$. Let x be a $\mathbb{M}$-superficial element and write $\overline{\mathbb{M}} = \mathbb{M}/x\mathbb{M}$. Now $M_2 : x \neq M_1$ since $v_1(\mathbb{M}) > 1 = v_1(\overline{\mathbb{M}})$.

Moreover we know that $e_2(\mathbb{M}) = 0$ and $e_2(\overline{\mathbb{M}}) = 1$ hence, by Proposition 1.2(4), we get $\sum_{i\geq 0} \lambda(M_{i+1} : x/M_i) = 1$ and then $M_{i+1} : x = M_i$ for every $i \geq 2$. As a consequence it follows that $\widetilde{M}_i = M_i$ for every $i \geq 2$, as wanted. Since

$$P_{\mathbb{M}}(z) = P_{\widetilde{\mathbb{M}}}(z) + \sum_{i\geq 0} [\lambda(\widetilde{M}_{i+1}/M_{i+1}) - \lambda(\widetilde{M}_i/M_i)]z^i,$$

by (4.2) and the previous fact, we get

$$\begin{aligned} P_{\mathbb{M}}(z) &= \frac{h_0(\widetilde{\mathbb{M}}) + (e_0(\mathbb{M}) - h_0(\widetilde{\mathbb{M}}))z}{(1-z)^2} + (1-z) = \\ &= \frac{h_0(\mathbb{M}) + (e_0(\mathbb{M}) - h_0(\mathbb{M}) - 2)z + 3z^2 - z^3}{(1-z)^2}. \end{aligned} \tag{4.3}$$

$\square$

Examples 3.3 and 3.4 show that in Theorem 4.5 both the Hilbert series can occur. We remark that Theorem 4.5 cannot be extended to dimension ≥ 3. In fact if we consider in $R = k[[x,y,z]]$ the q-adic filtration with $\mathfrak{q} = (x^2 - y^2, x^2 - z^2, xy, xz, yz)$, then $h_0(\mathfrak{q}) = \lambda(R/\mathfrak{q}) = 5$, $e_0(\mathfrak{q}) = 8$, $e_1(\mathfrak{q}) = 4 = e_0(\mathfrak{q}) - h_0(\mathfrak{q}) + 1$, but

$$P_{\mathfrak{q}}(z) = \frac{5 + 6z^2 - 4z^3 + z^4}{(1-z)^3.}$$

In this case depth $gr_{\mathfrak{q}}(R) = 0$ because $x^2 \notin \mathfrak{q}$ and $x^2 \in \mathfrak{q}^2 : \mathfrak{q}$.

In the last section we will characterize all the possible Hilbert series in higher dimension by using a recent result by Goto, Nishida, Ozeki on the structure of Sally modules of an m-primary ideal q satisfying the equality $e_1(\mathfrak{q}) = e_0(\mathfrak{q}) - \lambda(A/\mathfrak{q}) + 1$ (see [33]).

It is clear that, by using Sally's machine, the above Theorem 4.5 has a natural extension to higher dimensions.

Corollary 4.5. *Let M be a Cohen–Macaulay module of dimension $r \geq 2$, L a submodule of M and let $\mathbb{M} = \mathbb{M}_L$ be the good $\mathfrak{q}$-filtration on M induced by L. Assume $e_1(\mathbb{M}) = e_0(\mathbb{M}) - h_0(\mathbb{M}) + 1$ and depth $gr_\mathbb{M}(M) \geq r - 2$, then either*

$$P_\mathbb{M}(z) = \frac{h_0(\mathbb{M}) + h_1(\mathbb{M})z + z^2}{(1-z)^r}$$

or

$$P_\mathbb{M}(z) = \frac{h_0(\mathbb{M}) + h_1(\mathbb{M})z + 3z^2 - z^3}{(1-z)^r}.$$

4.4 The Case $e_1(\mathbf{M}) = e_0(\mathbf{M}) - h_0(\mathbf{M}) + 2$

In the case where $e_1(\mathbb{M}) = e_0(\mathbb{M}) - h_0(\mathbb{M}) + 2$ we present only partial results. The problem is open if $M_2 \cap JM \neq JM_1$ and $e_2 \neq 2$.

Theorem 4.6. *Let M be a Cohen–Macaulay module of dimension $r \geq 2$, L a submodule of M and let $\mathbb{M} = \mathbb{M}_L$ be the good $\mathfrak{q}$-filtration on M induced by L. Assume that $e_1(\mathbb{M}) = e_0(\mathbb{M}) - h_0(\mathbb{M}) + 2$ and $M_2 \cap JM = JM_1$ where J is an ideal generated by a maximal $\mathbb{M}$-superficial sequence for $\mathfrak{q}$. Then either $gr_\mathbb{M}(M)$ is Cohen–Macaulay and*

$$P_\mathbb{M}(z) = \frac{h_0(\mathbb{M}) + h_1(\mathbb{M})z + 2z^2}{(1-z)^r},$$

or depth $gr_\mathbb{M}(M) = r - 1$ and

$$P_\mathbb{M}(z) = \frac{h_0(\mathbb{M}) + h_1(\mathbb{M})z + z^3}{(1-z)^r}.$$

Proof. Since the assumptions are preserved modulo superficial elements in J, we may assume $r = 2$. By Theorem 4.3 we have

$$r_J(\mathbb{M}) \leq 3;$$

if $r(\mathbb{M}) \leq 2$ then $v_j(\mathbb{M}) = 0$ for every $j \geq 2$ so that $vv_j(\mathbb{M}) = 0$ for the same values of j. Since by assumption $vv_1(\mathbb{M}) = 0$, the associated graded module $gr_\mathbb{M}(M)$ is Cohen–Macaulay. Thus, by Theorem 2.5, we get

$$e_0(\mathbb{M}) - h_0(\mathbb{M}) + 2 = e_1(\mathbb{M}) = v_0(\mathbb{M}) + v_1(\mathbb{M}),$$

which implies $v_1(\mathbb{M}) = 2$, $e_2(\mathbb{M}) = 2$ and $e_j(\mathbb{M}) = 0$ for every $j \geq 3$. These values of the e_i's give the required Hilbert series.

Let $r(\mathbb{M}) = 3$; we have

$$M_2 \cap J\widetilde{M}_1 \subseteq M_2 \cap JM = JM_1$$

so that $1 \in S_J$. Further, depth $gr_{\widetilde{\mathbb{M}}}(M) \geq 1 = r - 1$, hence

$$e_1(\mathbb{M}) = e_1(\widetilde{\mathbb{M}}) = \sum_{i \geq 0} v_i(\widetilde{M}).$$

By Theorem 4.2 we get

$$\begin{aligned}
3 = r(\mathbb{M}) &\leq \sum_{i \geq 0} v_i(\widetilde{M}) + 2 - v_0(\mathbb{M}) - v_1(\mathbb{M}) \\
&= e_1(\mathbb{M}) + 2 - v_0(\mathbb{M}) - v_1(\mathbb{M}) \\
&= e_0(\mathbb{M}) - h_0(\mathbb{M}) + 2 + 2 - e_0(\mathbb{M}) + h_0(\mathbb{M}) - v_1(\mathbb{M}) \\
&= 4 - v_1(\mathbb{M})
\end{aligned}$$

which implies $v_1(\mathbb{M}) \leq 1$. Since $r(\mathbb{M}) = 3$, we cannot have $v_1(\mathbb{M}) = 0$, hence $v_1(\mathbb{M}) = 1$ and, by Corollary 4.4, we get

$$P_{\mathbb{M}}(z) = \frac{h_0(\mathbb{M}) + h_1(\mathbb{M})z + z^s}{(1 - z)^2}.$$

This gives

$$h_1(\mathbb{M}) + s = e_1(\mathbb{M}) = e_0(\mathbb{M}) - h_0(\mathbb{M}) + 2 = h_0(\mathbb{M}) + h_1(\mathbb{M}) + 1 - h_0(\mathbb{M}) + 2,$$

which implies $s = 3$ and depth $gr_{\mathbb{M}}(M) = 1$.   □

Let us remark that, as already noted in [25] for the case of the m-adic filtration, both the Hilbert functions given in the theorem are realizable.

In the above theorem we can get rid of the assumption $M_2 \cap JM = JM_1$, but then we need another strong requirement, the condition $e_2(\mathbb{M}) = 2$. The following theorem has been proved in the case of q-adic filtration by Sally (see [93]). This is a very deep result which gives a new class of ideals for which there is equality of the Hilbert function $H_I(n)$ and the Hilbert polynomial $p_I(n)$ at $n = 1$.

Theorem 4.7. *Let M be a Cohen–Macaulay module of dimension $r \geq 2$, L a submodule of M and let $\mathbb{M} = \mathbb{M}_L$ be the q-good filtration on M induced by L. Assume that $e_1(\mathbb{M}) = e_0(\mathbb{M}) - h_0(\mathbb{M}) + 2$ and $e_2(\mathbb{M}) = 2$. Then depth $gr_{\mathbb{M}}(M) \geq r - 1$ and*

$$P_{\mathbb{M}}(z) = \frac{h_0(\mathbb{M}) + h_1(\mathbb{M})z + 2z^2}{(1 - z)^r}.$$

Proof. By Sally's machine, we may reduce the problem to the case $r = 2$. As usual, J is an ideal generated by a maximal sequence of $\mathbb{M}$-superficial elements for q. First of all, we show that $M_1 = \tilde{M}_1$. Let

$$\tilde{L} := \tilde{M}_1 = M_{k+1} : q^k = q^k L : q^k$$

and $\mathbb{N}$ be the q-good filtration on M induced by $\tilde{L}$, so that

$$\tilde{\mathbb{M}} = \{M \supseteq \tilde{M}_1 \supseteq \tilde{M}_2 \supseteq \cdots \supseteq \tilde{M}_{j+1} \supseteq \cdots\}$$

where $\tilde{M}_{j+1} = \bigcup_k (q^{k+j}L : q^k)$ and

$$\mathbb{N} := \{M \supseteq \tilde{L} \supseteq q\tilde{L} \supseteq \cdots \supseteq q^j\tilde{L} \supseteq \cdots\}.$$

If $a \in q^j$ and $m \in \tilde{L}$, then we have

$$amq^k \subseteq aq^k L \subseteq q^{j+k}L.$$

Hence

$$M_{j+1} = q^j L \subseteq N_{j+1} = q^j \tilde{L} \subseteq q^{j+k}L : q^k \subseteq \tilde{M}_{j+1}.$$

This implies that
$$e_i(\mathbb{M}) = e_i(\mathbb{N}) = e_i(\tilde{\mathbb{M}})$$

for every $i = 0, \cdots, r$.
We apply (2.16) to the filtration $\mathbb{N}$ and we get

$$e_1(\mathbb{N}) = e_1(\mathbb{M}) \geq e_0(\mathbb{N}) - h_0(\mathbb{N}) = e_0(\mathbb{M}) - h_0(\mathbb{N}).$$

By Theorems 2.9 and 3.3 and since $e_2(\mathbb{M}) = e_2(\mathbb{N}) = 2$, we get

$$e_1(\mathbb{M}) \geq e_0(\mathbb{M}) - h_0(\mathbb{N}) + 2.$$

So we have

$$\lambda(M/\tilde{M}_1) = h_0(\mathbb{N}) \geq e_0(\mathbb{M}) - e_1(\mathbb{M}) + 2 = h_0(\mathbb{M}) = \lambda(M/M_1)$$

which implies $M_1 = \tilde{M}_1$.
Since $\operatorname{depth} gr_{\tilde{\mathbb{M}}}(M) \geq 1 = r - 1$, we have

$$e_2(\mathbb{M}) = e_2(\tilde{\mathbb{M}}) = \sum_{j \geq 1} jv_j(\tilde{\mathbb{M}}) = 2.$$

If $v_1(\widetilde{\mathbb{M}}) = 0$, then $M_2 \subseteq \widetilde{M}_2 = J\widetilde{M}_1 = JM_1 \subseteq M_2$ and therefore, by Corollary 2.6, $e_2(\mathbb{M}) = 0$, a contradiction.

Thus $v_1(\widetilde{\mathbb{M}}) = 2$, and $v_j(\widetilde{\mathbb{M}}) = 0$ for every $j \geq 2$. From this we get

$$2 = \lambda(\widetilde{M}_2/JM_1) = \lambda(\widetilde{M}_2/M_2) + \lambda(M_2/JM_1). \tag{4.4}$$

We cannot have $\lambda(M_2/JM_1) = v_1(\mathbb{M}) \leq 1$, otherwise $e_2(\mathbb{M}) \neq 2$ by Corollaries 2.6 and 4.4

Hence, we must have $v_1(\mathbb{M}) = 2$, so that, by (4.4), $\widetilde{M}_2 = M_2$. Since $v_j(\widetilde{\mathbb{M}}) = 0$ for every $j \geq 2$, we immediately get $\widetilde{M}_j = M_j$ for every $j \geq 0$ and depth$gr_{\mathbb{M}}(M) \geq 1$. By Theorem 2.5, this implies $e_2(\mathbb{M}) = \sum_{j \geq 1} j v_j(\mathbb{M}) = 2$, hence $v_j(\mathbb{M}) = 0$ for every $j \geq 2$ and $e_j(\mathbb{M}) = 0$ for every $j \geq 3$; this gives the required Hilbert series.   □

The following example shows that the assumptions $e_1(\mathbb{M}) = e_0(\mathbb{M}) - h_0(\mathbb{M}) + 2$ and $e_2(\mathbb{M}) = 2$ in Theorem 4.7 do not imply that $gr_{\mathbb{M}}(M)$ is Cohen–Macaulay.

Example 4.1. Let $A = k[[x, y]]$, $q = (x^6, x^5 y, x^4 y^9, x^3 y^{15}, x^2 y^{16}, xy^{22}, y^{24})$ and $\mathbb{M}$ the q-adic filtration. Then we have

$$P_{\mathbb{M}}(z) = \frac{87 + 37z + 2z^2}{(1-z)^2}$$

so that $e_0(\mathbb{M}) = 126$, $e_1(\mathbb{M}) = 41$, $e_2(\mathbb{M}) = 2$, and $h_0(\mathbb{M}) = 87$. We have $41 = 126 - 87 + 2$ but the associated graded ring is not Cohen–Macaulay.

The following example shows that, in Theorem 4.7, the assumption $e_2 = 2$ is essential.

Example 4.2. Let $A = k[[x, y]]$, $q = (x^6, x^5 y^2, x^4 y^6, x^3 y^8, x^2 y^9, xy^{11}, y^{13})$ and $\mathbb{M}$ the q-adic filtration. Then we have

$$P_{\mathbb{M}}(z) = \frac{49 + 25z + 3z^2 + z^3 - z^4}{(1-z)^2}$$

so that $e_0(\mathbb{M}) = 77$, $e_1(\mathbb{M}) = 30$ and $h_0(\mathbb{M}) = 49$. We have $30 = 77 - 49 + 2$. Further $x^4 y^5 \notin q$, $x^4 y^5 \in q^3 : q^2$, so that $M_1 \neq \widetilde{M}_1$. This implies that the associated graded ring has depth zero. Of course, $e_2(\mathbb{M}) = 0 \neq 2$.

Chapter 5
Applications to the Fiber Cone

Let $(A, \mathfrak{m})$ be a commutative local ring and let $\mathfrak{q}$ be an ideal of A. As usual $\mathbb{M}$ denotes a good $\mathfrak{q}$-filtration of a module M of dimension r and $gr_{\mathfrak{q}}(A) = \oplus_{n \geq 0} \mathfrak{q}^n / \mathfrak{q}^{n+1}$ the associated graded ring to $\mathfrak{q}$. Given an ideal I containing $\mathfrak{q}$, we define the graded module on $gr_{\mathfrak{q}}(A)$

$$F_I(\mathbb{M}) := \oplus_{n \geq 0} M_n / I M_n.$$

$F_I(\mathbb{M})$ is called the Fiber cone of $\mathbb{M}$ with respect to I. If $\mathbb{M}$ is the $\mathfrak{q}$-adic filtration on A and $I = \mathfrak{m}$, then we write $F_{\mathfrak{m}}(\mathfrak{q}) = \oplus_{n \geq 0} \mathfrak{q}^n / \mathfrak{m} \mathfrak{q}^n$ which is the classical definition of the Fiber cone of $\mathfrak{q}$. It coincides with $gr_{\mathfrak{m}}(A)$ when $\mathfrak{q} = \mathfrak{m}$. This graded object encodes a lot of information about $\mathfrak{q}$. For instance, its dimension gives the minimal number of generators of any minimal reduction of $\mathfrak{q}$, that is the analytic spread of $\mathfrak{q}$, and its Hilbert function determines the minimal number of generators of the powers of $\mathfrak{q}$. We remark that if $F_{\mathfrak{m}}(\mathfrak{q})$ is Cohen–Macaulay, then the reduction number of $\mathfrak{q}$ can be read off directly from the Hilbert series of $F_{\mathfrak{m}}(\mathfrak{q})$ (see [110, Proposition 1.85]). Usually the arithmetical properties of the Fiber cone and those of the associated graded ring were studied via apparently different approaches. The literature concerning the associated graded rings is much richer, but new and specific techniques were necessary in order to study the Fiber cone. In spite of the fact that $F_I(\mathbb{M})$ is not the graded module associated to a filtration, the aim of this section is to show that it is possible to deduce information about $F_I(\mathbb{M})$ as a consequence of the theory on filtrations. In particular we will obtain recent results on the Fiber cone of an ideal as an easy consequence of classical results on the associated graded rings of certain special filtrations. In this chapter we prove some results which were recently obtained using different devices, often very technical ones. Of course, we are not going to give a complete picture of the literature on the Fiber Cone, but we have selected some results which illustrate well the use of this approach. Our hope is that this method will be useful in the future to prove new results on this topic.

5.1 Depth of the Fiber Cone

Cortadellas and Zarzuela proved in [18] and [17] the existence of an exact sequence of the homology of modified Koszul complexes which relates $F_I(\mathbb{M})$ with the associated graded modules to the filtrations $\mathbb{M}$ and $\mathbb{M}^I$. We recall that the filtration $\mathbb{M}^I$ on M is defined as follows:

$$\mathbb{M}^I: \quad M \supseteq IM \supseteq IM_1 \supseteq \cdots \supseteq IM_n \cdots \supseteq \ldots \tag{5.1}$$

It is clear that $\mathbb{M}^I$ is a good q-filtration on M, thus $e_0(\mathbb{M}) = e_0(\mathbb{M}^I)$.

Starting from their work, we are going to present several applications using the results on the filtered modules which have been presented in the previous chapters. First we prove a result which relates the depth of $F_I(\mathbb{M})$, with the depths of $gr_{\mathbb{M}}(M)$ and $gr_{\mathbb{M}^I}(M)$. Since the involved objects are graded modules on $gr_{\mathfrak{q}}(A)$, the depths are always computed with respect to $Q = \oplus_{n>0}\mathfrak{q}^n/\mathfrak{q}^{n+1}$.

Proposition 5.1. *Let $\mathbb{M}$ be a good q-filtration on a module M and let I be an ideal containing q such that $M_{n+1} \subseteq IM_n$ for every $n \geq 1$. We have:*

(1) depth $F_I(\mathbb{M}) \geq min\{$depth $gr_{\mathbb{M}}(M) + 1,$ depth $gr_{\mathbb{M}^I}(M)\}$
(2) depth $gr_{\mathbb{M}^I}(M) \geq min\{$depth $gr_{\mathbb{M}}(M),$ depth $F_I(\mathbb{M})\}$
(3) depth $gr_{\mathbb{M}}(M) \geq min\{$depth $gr_{\mathbb{M}^I}(M),$ depth $F_I(\mathbb{M})\} - 1.$

Proof. We have the following homogeneous exact sequences of $gr_{\mathfrak{q}}(A)$-graded modules:

$$0 \to N \to gr_{\mathbb{M}}(M) \longrightarrow F_I(\mathbb{M}) \to 0$$

$$0 \to F_I(\mathbb{M}) \to gr_{\mathbb{M}^I}(M) \longrightarrow N(-1) \to 0$$

where $N = \oplus_{n \geq 0} IM_n/M_{n+1}$.

It is enough to remark that we have the following exact sequences of the corresponding homogeneous parts of degree n :

$$0 \to IM_n/M_{n+1} \to M_n/M_{n+1} \to M_n/IM_n \to 0$$

$$0 \to M_n/IM_n \to IM_{n-1}/IM_n \longrightarrow IM_{n-1}/M_n \to 0.$$

The inequality between the depths follows from standard facts (see for example the depth formula in [6]). $\qquad\qquad\Box$

Several examples show that $F_{\mathfrak{m}}(\mathfrak{q})$ can be Cohen–Macaulay even if $gr_{\mathfrak{q}}(A)$ is not Cohen–Macaulay and conversely. The above proposition clarifies the intermediate role of the graded module associated to the filtration $\mathbb{M} = \{\mathfrak{m}\mathfrak{q}^n\}$.

It will be useful to remind that, by Remark 1.1, it is possible to find a superficial sequence $a_1, \ldots, a_r$ in q which is both $\mathbb{M}$-superficial and $\mathbb{M}^I$-superficial for q.

The above proposition gives freely several recent results proved in the literature with different and heavy methods. We can quote as examples Theorems 1 and 2 in [96], Theorem 2 in [87], Theorem 3.4 in [55], Proposition 4.1, Corollary 4.3, Proposition 4.4 in [17].

We present here a proof of Theorem 1 in [96], to show an explicit application of our approach. We prove the original statement, but it could easily be extended to modules.

Theorem 5.1. *Let $\mathfrak{q}$ be an ideal of a local ring $(A, \mathfrak{m})$ and let J be an ideal generated by a superficial regular sequence for $\mathfrak{q}$ such that $\mathfrak{q}^2 = J\mathfrak{q}$. Then $F_\mathfrak{m}(\mathfrak{q})$ is Cohen–Macaulay.*

Proof. By using the assumption, we get that $\mathfrak{q}^{n+1} \cap J = J\mathfrak{q}^n$ and $\mathfrak{m}\mathfrak{q}^{n+1} \cap J = J\mathfrak{m}\mathfrak{q}^n$ for every integer n. By the Valabrega–Valla criterion it follows that the filtrations $\{\mathfrak{q}^n\}$ and $\{\mathfrak{m}\mathfrak{q}^n\}$ on A have associated graded rings of depth at least $\mu(J) = \dim F_\mathfrak{m}(\mathfrak{q})$. The result follows now by Proposition 5.1(1).   □

We remark that in the above case we can easily write the Hilbert series of the standard graded k-algebra $F_\mathfrak{m}(\mathfrak{q})$, that is $P_{F_\mathfrak{m}(\mathfrak{q})}(z) = \sum_{i\geq 0} \dim_k(\mathfrak{q}^i/\mathfrak{m}\mathfrak{q}^i)z^i$. In fact

$$P_{F_\mathfrak{m}(\mathfrak{q})}(z) = \frac{1}{(1-z)^r} P_{F_\mathfrak{m}(\mathfrak{q})/JF_\mathfrak{m}(\mathfrak{q})}(z) = \frac{1}{(1-z)^r} \sum_{i\geq 0}^{s} \dim_k(\mathfrak{q}^i/J\mathfrak{q}^{i-1} + \mathfrak{m}\mathfrak{q}^i)z^i$$

Since $\mathfrak{q}^2 = J\mathfrak{q}$ and $\dim_k(\mathfrak{q}/J + \mathfrak{m}\mathfrak{q}) = \mu(\mathfrak{q}) - r$, one has

$$P_{F_\mathfrak{m}(\mathfrak{q})}(z) = \frac{1 + (\mu(\mathfrak{q}) - r)z}{(1-z)^r}.$$

5.2   The Hilbert Function of the Fiber Cone

If $\lambda(M/IM)$ is finite, then M_n/IM_n has finite length and we may define for every integer n the numerical function

$$H_{F_I(\mathbb{M})}(n) := \lambda(M_n/IM_n)$$

which is the Hilbert function of $F_I(\mathbb{M})$. We denote by $P_{F_I}(z)$ the corresponding Hilbert series, that is $\sum_{i\geq 0} H_{F_I(\mathbb{M})}(i)z^i$.

From now on we shall assume that $\lambda(M/\mathfrak{q}M)$ is finite. In this case $\dim F_I(\mathbb{M}) = r = \dim M$. We recall that $H_{F_I(\mathbb{M})}(n)$ is a polynomial function and the corresponding polynomial $p_{F_I}(X)$ has degree $r - 1$. It is the Hilbert polynomial of $F_I(\mathbb{M})$ and, as usual, we can write

$$p_{F_I(\mathbb{M})}(X) = \sum_{i=0}^{r-1} (-1)^i f_i(\mathbb{M}) \binom{X + r - i - 1}{r - i - 1}.$$

The coefficients $f_i(\mathbb{M})$ are integers and they are called the Hilbert coefficients of $F_I(\mathbb{M})$. In particular $f_0(\mathbb{M})$ is the multiplicity of the fiber cone of $\mathbb{M}$.

We can relate the Hilbert coefficients of $F_I(\mathbb{M})$ to those of the filtrations $\mathbb{M}$ and $\mathbb{M}^I$ in a natural way. We remark that, for every $n \geq 0$, we have

$$\lambda(M/M_n) + \lambda(M_n/IM_n) = \lambda(M/IM_n) \tag{5.2}$$

Hence

$$p_{\mathbb{M}}^1(X-1) + p_{F_I}(X) = p_{\mathbb{M}^I}^1(X) \tag{5.3}$$

and

$$zP_{\mathbb{M}}^1(z) + P_{F_I}(z) = P_{\mathbb{M}^I}^1(z). \tag{5.4}$$

Since

$$p_{\mathbb{M}}^1(X-1) = \sum_{i=0}^{r}(-1)^i e_i(\mathbb{M})\binom{X+r-i-1}{r-i}$$

and

$$p_{\mathbb{M}^I}^1(X) = \sum_{i=0}^{r}(-1)^i e_i(\mathbb{M}^I)\binom{X+r-i}{r-i}$$

from (5.3), it is possible to prove that

$$e_0(\mathbb{M}) = e_0(\mathbb{M}^I) \quad \text{and} \quad f_{i-1}(\mathbb{M}) = e_i(\mathbb{M}) + e_{i-1}(\mathbb{M}) - e_i(\mathbb{M}^I) \tag{5.5}$$

for every $i = 1, \ldots, r$.

Hence the theory developed in the previous sections on the Hilbert coefficients of the graded module associated to a good filtration on M can be applied to $e_i(\mathbb{M})$ and $e_i(\mathbb{M}^I)$ in order to get information, via (5.5), on the coefficients of the Fiber cone of $\mathbb{M}$.

5.3 A Version of Sally's Conjecture for the Fiber Cone

We present a short proof of the main result of [43, Theorem 4.4], which is the analog of Sally's conjecture in the case of the fiber cone.

Theorem 5.2. *Let $\mathbb{M}$ be the $\mathfrak{q}$-adic filtration on a Cohen–Macaulay module M of dimension r and let I be an ideal containing $\mathfrak{q}$. Assume $\mathbb{M}$ has almost Goto minimal multiplicity with respect to I and depth $gr_{\mathbb{M}}(M) \geq r - 2$.*
Then depth $F_I(\mathbb{M}) \geq r - 1$.

Proof. We recall that $\mathbb{M}$ has almost Goto minimal multiplicity with respect to I if and only if $\mathbb{M}^I$ has almost minimal multiplicity if and only if one has $\lambda(IM_1/JIM) = 1$ for every ideal J generated by a maximal superficial sequence for $\mathfrak{q}$, equivalently $\lambda(I\mathfrak{q}M/JIM) = 1$. Hence by Corollary 4.4, we get depth $gr_{\mathbb{M}^I}(M) \geq r - 1$ and the result follows now by Proposition 5.1. $\square$

We remark that, under the assumptions of Theorem 5.2, we are able to write the Hilbert series of $F_I(\mathbb{M})$. In fact, by using (5.4) and Theorem 4.4, we get

$$P_{F_I(\mathbb{M})}(z) = \frac{\lambda(M/IM) + [e_0((\mathbb{M}) - \lambda(M/IM) - 1]z + z^s - zh_{\mathbb{M}}(z)}{(1-z)^{r+1}}$$

for some integer $s \geq 2$.

The following example shows that in Theorem 5.2, the assumption depth $gr_{\mathbb{M}}(M) \geq r - 2$ is necessary.

Example 5.1. Let $A = k[[x,y,z]]$ and $q = (y^2 - x^2, z^2 - y^2, xy, yz, zx)$. The ideal $J = (y^2 - x^2, z^2 - y^2, xy)$ is generated by a maximal superficial sequence for q and $\lambda(mq/mJ) = 1$. Therefore if we consider $\mathbb{M} = \{q^n\}$ the q-adic filtration on A and $I = m$ the maximal ideal of A, the filtration $\mathbb{M}$ has Goto almost minimal multiplicity since $\mathbb{M}^I = \{mq^n\}$ has almost minimal multiplicity. Since $x^2 \in q^2 : q$, but $x^2 \notin q$, it follows that depth $gr_q(A) = 0$. In this case depth $F_q(A) = 1 < r - 1$ (cf. Example 4.5 in [43]).

It is possible to prove the above theorem as consequence of the following more general result.

Theorem 5.3. *Let $\mathbb{M}$ be the q-adic filtration on a Cohen–Macaulay module M of dimension r and let I be an ideal containing q. Assume:*

(1) depth $gr_{\mathbb{M}}(M) \geq r - 2$

(2) $\lambda(Iq^2M/JIqM) \leq 1$ *and* $IqM \cap JM = IJM$ *for some ideal J generated by a maximal superficial sequence for $\mathbb{M}^I$.*

Then depth $\Gamma_I(\mathbb{M}) \geq r - 1$.

Proof. Since $IqM \cap JM = IJM$ and we assume $\lambda(Iq^2M/JIqM) \leq 1$, by Theorem 4.4 applied to $\mathbb{M}^I$, we get depth $gr_{\mathbb{M}^I}(M) \geq r - 1$. The result follows now by Proposition 5.1. $\qquad\qquad\square$

We remark that in the classical case of the q-adic filtration of A and $I = m$, the assumption $mq \cap J = mJ$ is always satisfied.

In Theorem 5.2 we discussed depth $F_I(\mathbb{M})$ when $\mathbb{M}^I$ has almost minimal multiplicity. A natural question arises about the depth of $F_I(\mathbb{M})$ when $\mathbb{M}$ has almost minimal multiplicity, that is

$$e_0(\mathbb{M}) = (1 - r)\lambda(M/M_1) + \lambda(M_1/M_2) + 1$$

or equivalently $\lambda(M_2/JM_1) = 1$ for every ideal J generated by a maximal $\mathbb{M}$-superficial sequence. By Corollary 4.4, this assumption guarantees that depth $gr_{\mathbb{M}}(M) \geq r - 1$. Examples show that $F_I(\mathbb{M})$ is not necessarily Cohen–Macaulay and it is natural to ask whether depth $F_I(\mathbb{M}) \geq r - 1$.

We have the analogous result of Theorem 5.2 where the proof works in the same way essentially swapping $\mathbb{M}^I$ with $\mathbb{M}$.

Theorem 5.4. *Let* $\mathbb{M}$ *be the* q-*adic filtration on a Cohen–Macaulay module* M *of dimension* r *and let* I *be an ideal containing* q. *Assume* $\mathbb{M}$ *has almost minimal multiplicity and* depth $gr_{\mathbb{M}^I}(M) \geq r-1$.
    Then depth $F_I(\mathbb{M}) \geq r-1$.

In a recent paper, A.V. Jayanthan, T. Puthenpurakal and J. Verma [40, Theorem 3.4] proved a criterion for the Cohen–Macaulayness of $F_{\mathfrak{m}}(\mathfrak{q})$ when q has almost minimal multiplicity giving an answer to a question raised by G.Valla. We give here a proof by using our approach.

Theorem 5.5. *Let* q *be an* $\mathfrak{m}$-*primary ideal of a local Cohen–Macaulay ring* $(A,\mathfrak{m})$ *of dimension* r. *Assume* q *has almost minimal multiplicity and let* J *be an ideal generated by a maximal superficial sequence for* q. *Then the following conditions are equivalent:*

(1) $\mathfrak{m}\mathfrak{q}^2 = J\mathfrak{m}\mathfrak{q}$

(2) $F_{\mathfrak{m}}(\mathfrak{q})$ *is Cohen–Macaulay*

(3) $P_{F_{\mathfrak{m}}(\mathfrak{q})}(z) = \frac{1+\lambda(\mathfrak{q}/J+\mathfrak{q}\mathfrak{m})z+z^2+\cdots+z^s}{(1-z)^r}$ *for some integer* $s \geq 2$.

Proof. As usual, denote by $\mathbb{M}^{\mathfrak{m}} = \{\mathfrak{m}\mathfrak{q}^n\}$ the filtration on A. Since q has almost minimal multiplicity, then $\lambda(\mathfrak{q}^2/J\mathfrak{q}) = 1$. Hence, by Corollary 4.4, depth $gr_{\mathfrak{q}}(A) \geq r-1$. Now, if $\mathfrak{m}\mathfrak{q}^2 = J\mathfrak{m}\mathfrak{q}$, by using the Valabrega–Valla criterion, we have that $gr_{\mathbb{M}^{\mathfrak{m}}}(A)$ is Cohen–Macaulay and hence $F_{\mathfrak{m}}(\mathfrak{q})$ is Cohen–Macaulay by Proposition 5.1 proving (1) implies (2).

Assume now that $F_{\mathfrak{m}}(\mathfrak{q})$ is Cohen–Macaulay. If $J = (a_1, \ldots, a_r)$, we recall that the corresponding classes in $\mathfrak{q}/\mathfrak{m}\mathfrak{q}$ form a system of parameters for $F_{\mathfrak{m}}(\mathfrak{q})$ and hence a regular sequence on $F_{\mathfrak{m}}(\mathfrak{q})$. It follows that

$$P_{F_{\mathfrak{m}}(\mathfrak{q})}(z) = \frac{1}{(1-z)^r}P_{F_{\mathfrak{m}}(\mathfrak{q})/JF_{\mathfrak{m}}(\mathfrak{q})}(z) = \frac{1}{(1-z)^r}\sum_{i\geq 0}\lambda(\mathfrak{q}^i/J\mathfrak{q}^{i-1}+\mathfrak{m}\mathfrak{q}^i)z^i.$$

Since $\lambda(\mathfrak{q}^2/J\mathfrak{q}) = 1$, then $\lambda(\mathfrak{q}^{i+1}/J\mathfrak{q}^i) \leq 1$ for every $i \geq 1$. Let $s \geq 2$ the least integer such that $\mathfrak{q}^{s+1} = J\mathfrak{q}^s$. Since $\lambda(\mathfrak{q}^i/J\mathfrak{q}^{i-1}) = 1$ for $i = 2, \ldots, s$ and hence $\mathfrak{m}\mathfrak{q}^i \subseteq J\mathfrak{q}^{i-1}$ for every $i \geq 2$, we get

$$P_{F_{\mathfrak{m}}(\mathfrak{q})}(z) = \frac{1+\lambda(\mathfrak{q}/J+\mathfrak{q}\mathfrak{m})z+z^2+\cdots+z^s}{(1-z)^r}.$$

It follows that (2) implies (3). Actually (2) is equivalent to (3) In fact if $P_{F_{\mathfrak{m}}(\mathfrak{q})}(z) = \frac{1}{(1-z)^r}P_{F_{\mathfrak{m}}(\mathfrak{q})/JF_{\mathfrak{m}}(\mathfrak{q})}(z)$, then $F_{\mathfrak{m}}(\mathfrak{q})$ is Cohen–Macaulay.

We prove now (3) implies (1). We consider the filtration $\mathbb{M}_{\mathfrak{m}}$ defined in (2.18) and we have to prove that is $v_2(\mathbb{M}_{\mathfrak{m}}) = \lambda(\mathfrak{m}\mathfrak{q}^2/J\mathfrak{m}\mathfrak{q}) = 0$. Since $F_{\mathfrak{m}}(\mathfrak{q})$ is Cohen–Macaulay and depth $gr_{\mathfrak{q}}(A) \geq r-1$, by Proposition 5.1, depth $gr_{\mathbb{M}_{\mathfrak{m}}}(A) \geq r-1$. Hence, by Theorem 2.5, $e_1(\mathbb{M}_{\mathfrak{m}}) = \sum_{i\geq 0}v_i(\mathbb{M}_{\mathfrak{m}})$. Then we have to prove $e_1(\mathbb{M}_{\mathfrak{m}}) = v_0(\mathbb{M}_{\mathfrak{m}}) + v_1(\mathbb{M}_{\mathfrak{m}}) = e_0(\mathbb{M}_{\mathfrak{m}}) - 1 + \lambda(\mathfrak{m}\mathfrak{q}/J\mathfrak{m})$.

Now, by (5.5), we know that $e_1(\mathbb{M}_{\mathrm{m}}) = e_0(\mathbb{M}_{\mathrm{m}}) + e_1(\mathfrak{q}) - f_0(\mathfrak{q})$. Since $e_1(\mathfrak{q}) = $ $= \sum_{i \geq s} v_i(\mathfrak{q}) = \lambda(\mathfrak{q}/J) + s - 1$ and $f_0 = 1 + \lambda(\mathfrak{q}/J + \mathfrak{q}\mathrm{m}) + s - 1$, we have $e_1(\mathbb{M}_{\mathrm{m}}) = $ $= e_0(\mathbb{M}_{\mathrm{m}}) + \lambda(\mathfrak{q}/J) + s - 1 - (1 + \lambda(\mathfrak{q}/J + \mathfrak{q}\mathrm{m}) + s - 1) = e_0(\mathbb{M}_{\mathrm{m}}) - 1 + \lambda(\mathrm{m}\mathfrak{q}/J\mathrm{m})$, as required. $\square$

5.4 The Hilbert Coefficients of the Fiber Cone

The formulas in (5.5) give information on the Hilbert coefficients of the Fiber Cone by means of the theory of Hilbert functions of filtered modules. First we get a short proof of a recent result by A. Corso (see [12]).

Theorem 5.6. *Let $\mathbb{M}$ be a good $\mathfrak{q}$-filtration on a module M and let I be an ideal containing $\mathfrak{q}$ such that $M_{n+1} \subseteq IM_n$. Let J be the ideal generated by a maximal $\mathbb{M}^I$-superficial sequence for $\mathfrak{q}$ and denote by $\mathbb{N}$ the corresponding filtration $\{J^n M\}$. Then*

$$f_0(\mathbb{M}) \leq min\{e_1(\mathbb{M}) - e_0(\mathbb{M}) - e_1(\mathbb{N}) + \lambda(M/IM) + \lambda(M/IM_1 + JM),$$

$$e_1(\mathbb{M}) - e_1(\mathbb{N}) + \lambda(M/IM)\}.$$

Proof. Since $f_0(\mathbb{M}) = e_0(\mathbb{M}) + e_1(\mathbb{M}) - e_1(\mathbb{M}^{\mathbb{I}})$ by (5.5), it is enough to apply Theorem 2.4 to $e_1(\mathbb{M}^I)$ for $s = 1, 2$. $\square$

If we apply the above result to the case $\mathbb{M} = \{\mathfrak{q}^n\}$ with $I = \mathrm{m}$, we easily obtain the following bound on the multiplicity of the fiber cone $F_{\mathrm{m}}(\mathfrak{q})$ (see [12, Theorem 3.4]).

Corollary 5.1. *Let $\mathfrak{q}$ be an m-primary ideal of a local ring (A, m) of dimension r. Let J be the ideal generated by a maximal superficial sequence for $\mathfrak{q}$, then*

$$f_0(\mathfrak{q}) \leq min\{e_1(\mathfrak{q}) - e_0(\mathfrak{q}) - e_1(J) + \lambda(A/\mathfrak{q}) + \mu(\mathfrak{q}) - r + 1, e_1(\mathfrak{q}) - e_1(J) + 1\}.$$

If A is Cohen–Macaulay, then $e_1(J) = 0$ because J is generated by a regular sequence and we are able to characterize the extremal cases. The following result generalizes Proposition 2.2 and Theorem 2.5 in [15].

Corollary 5.2. *Let $\mathfrak{q}$ be an m-primary ideal of a local Cohen–Macaulay ring (A, m) of dimension r. Then*

$$f_0(\mathfrak{q}) \leq e_1(\mathfrak{q}) - e_0(\mathfrak{q}) + \lambda(A/\mathfrak{q}) + \mu(\mathfrak{q}) - r + 1 \leq e_1(\mathfrak{q}) + 1.$$

In particular:

(1) *If $f_0(\mathfrak{q}) = e_1(\mathfrak{q}) + 1$, then $\mathrm{m}\mathfrak{q} = \mathrm{m}J$ for every maximal superficial sequence J for $\mathfrak{q}$. If, in addition, $\lambda(\mathfrak{q}^2 \cap J/J\mathfrak{q}) \leq 1$ for some J, then $\mathrm{depth}\, gr_{\mathfrak{q}}(A) \geq r - 1$ and $F_{\mathrm{m}}(\mathfrak{q})$ is Cohen–Macaulay.*

(2) *If $f_0(\mathfrak{q}) = e_1(\mathfrak{q}) - e_0(\mathfrak{q}) + \lambda(A/\mathfrak{q}) + \mu(\mathfrak{q}) - r + 1$, then $F_{\mathrm{m}}(\mathfrak{q})$ is unmixed.*

Proof. The first inequality follows by Corollary 5.1. We prove now that $e_1(\mathfrak{q}) - e_0(\mathfrak{q}) + \lambda(A/\mathfrak{q}) + v(\mathfrak{q}) - r + 1 \le e_1(\mathfrak{q}) + 1$. If J is an ideal generated by a maximal superficial sequence for $\mathfrak{q}$, then $e_0(\mathfrak{q}) - \lambda(A/\mathfrak{q}) - v(\mathfrak{q}) + r = \lambda(\mathfrak{q}/J) - \lambda(\mathfrak{q}/\mathfrak{q}\mathfrak{m}) + \lambda(J/J\mathfrak{m}) = \lambda(\mathfrak{q}/J\mathfrak{m}) - \lambda(\mathfrak{q}/\mathfrak{q}\mathfrak{m}) \ge 0$.

In particular if $f_0(\mathfrak{q}) = e_1(\mathfrak{q}) + 1$, then $\mathfrak{q}\mathfrak{m} = J\mathfrak{m}$ and hence the associated graded module to the $\mathfrak{q}$-filtration $\mathbb{M}_\mathfrak{m} = \{\mathfrak{m}\mathfrak{q}^n\}$ is Cohen–Macaulay by the Valabrega–Valla criterion. Now $\mathfrak{q}^2 \subseteq \mathfrak{m}\mathfrak{q} = \mathfrak{m}J \subseteq J$, then $\lambda(\mathfrak{q}^2 \cap J/J\mathfrak{q}) = \lambda(\mathfrak{q}^2/J\mathfrak{q}) \le 1$. Hence by Theorem 4.4, depth $gr_\mathfrak{q}(A) \ge r - 1$ and (1) follows by Proposition 5.1.

Now, from the proof of Theorem 5.6, $f_0(\mathfrak{q}) = e_1(\mathfrak{q}) - e_0(\mathfrak{q}) + \lambda(A/\mathfrak{q}) + v(\mathfrak{q}) - r + 1$ if and only if $e_1(\mathbb{M}_\mathfrak{m}) = 2e_0(\mathbb{M}_\mathfrak{m}) - 1 - \lambda(A/\mathfrak{m}\mathfrak{q} + J) = 2e_0(\mathbb{M}_\mathfrak{m}) - 2h_0(\mathbb{M}_\mathfrak{m}) - h(\mathbb{M}_\mathfrak{m})$ and hence, by Theorem 2.8, $gr_{\mathbb{M}_\mathfrak{m}}(A)$ is Cohen–Macaulay. Because we have a canonical injective map from $F_\mathfrak{m}(\mathfrak{q})$ to $gr_{\mathbb{M}_\mathfrak{m}}(A)$ the result follows.   □

We remark that, in the above result, the assumption $\lambda(\mathfrak{q}^2 \cap J/J\mathfrak{q}) \le 1$ is satisfied for example if A is Gorenstein (see Proposition 2.2 in [12]).

5.5 Further Numerical Invariants: The g_i

We give now short proofs of several recent results proved in [43], [42] and [40]. First we need to relate the numerical invariants already considered to those introduced by A.V. Jayanthan and J. Verma. They write the polynomial $p^1_{\mathbb{M}^I}(X)$ of degree r by using the unusual binomial basis $\{\binom{X+r-i-1}{r-i} \ : \ i = 0, \dots, d\}$. The integers $g_i(\mathbb{M}^I)$ are uniquely determined

$$p^1_{\mathbb{M}^I}(X) = \sum_{i=0}^r (-1)^i g_i(\mathbb{M}^I) \binom{X+r-i-1}{r-i}.$$

They have the advantage of leading to more compact formulas than those in (5.5). By (5.3), it is easy to check that

$$e_0(\mathbb{M}) = g_0(\mathbb{M}^I) \quad \text{and} \quad g_i(\mathbb{M}^I) = e_i(\mathbb{M}) - f_{i-1}(\mathbb{M})$$

for every $i = 1, \dots, r$. Moreover from the following equalities

$$p^1_{\mathbb{M}^I}(X) = \sum_{i=0}^r (-1)^i g_i(\mathbb{M}^I) \binom{X+r-i-1}{r-i} =$$

$$= \sum_{i=0}^r (-1)^i e_i(\mathbb{M}^I) \binom{X+r-i}{r-i}$$

we obtain

$$e_0(\mathbb{M}^I) = g_0(\mathbb{M}^I) \quad \text{and} \quad e_i(\mathbb{M}^I) = g_{i-1}(\mathbb{M}^I) + g_i(\mathbb{M}^I)$$

for every $i = 1, \dots, r$. Then

$$g_i(\mathbb{M}^I) = \sum_{j=0}^{i} (-1)^{i-j} e_j(\mathbb{M}^I) \qquad (5.6)$$

In particular

$$g_1(\mathbb{M}^I) = e_1(\mathbb{M}^I) - e_0(\mathbb{M}^I) \quad \text{and} \quad g_2(\mathbb{M}^I) = e_2(\mathbb{M}^I) - e_1(\mathbb{M}^I) + e_0(\mathbb{M}^I).$$

From the equality (5.6) it is clear that the integers $g_i(\mathbb{M}^I)$ have good behaviour modulo $\mathbb{M}^I$-superficial elements for q (see [42, Lemma 3.5]).

With almost no further effort we can obtain and generalize several results in [42], among them Proposition 4.1 and Theorem 4.3. It will be useful to recall that we have $w_0(\mathbb{M}^I) = v_0(\mathbb{M}^I) = \lambda(M/JM) - \lambda(M/IM)$ and, for $n \geq 1$, $w_n(\mathbb{M}^I) = \lambda(IM_n + JM/JM)$ and $v_n(\mathbb{M}^I) = \lambda(IM_n/JIM_{n-1})$.

Theorem 5.7. *Let $\mathbb{M}$ be a good q-filtration of a Cohen–Macaulay module M of dimension r and let I be an ideal containing q such that $M_{n+1} \subseteq IM_n$. Let J be the ideal generated by a maximal $\mathbb{M}^I$-superficial sequence for q. Then*

$$g_1(\mathbb{M}^I) \geq \sum_{n \geq 1} w_n(\mathbb{M}^I) - \lambda(M/IM)$$

The equality holds if and only if $gr_{\mathbb{M}^I}(M)$ is Cohen–Macaulay.

Moreover if depth $gr_{\mathbb{M}}(M) \geq r - 1$, then $g_1(\mathbb{M}^I) = \sum_{n \geq 1} w_n(\mathbb{M}^I) - \lambda(M/IM)$ if and only if $F_I(\mathbb{M})$ is Cohen–Macaulay.

Proof. It is enough to recall that $g_1(\mathbb{M}^I) = e_1(\mathbb{M}^I) - e_0(\mathbb{M}^I)$. The result follows by Theorem 2.7 applied to the filtration $\mathbb{M}^I$. The last part is a consequence of the previous result and Proposition 5.1. $\qquad\square$

The next result extends and completes Theorem 2.5 (see also [42]).

Theorem 5.8. *Let $\mathbb{M}$ be a good q-filtration of a Cohen–Macaulay module M of dimension r and let I be an ideal containing q such that $M_{n+1} \subseteq IM_n$. Let J be the ideal generated by a maximal $\mathbb{M}^I$-superficial sequence for q. Then we have:*

(a) $g_1(\mathbb{M}^I) \leq \sum_{n \geq 1} v_n(\mathbb{M}^I) - \lambda(M/IM)$

(b) $g_2(\mathbb{M}^I) \leq \sum_{n \geq 2} (n-1) v_n(\mathbb{M}^I) + \lambda(M/IM)$

If depth $gr_{\mathbb{M}}(M) \geq r - 1$ the following conditions are equivalent:

(1) depth $F_I(\mathbb{M}) \geq r - 1$.

(2) $g_1(\mathbb{M}^I) = \sum_{n \geq 1} v_n(\mathbb{M}^I) - \lambda(M/IM)$

(3) $g_i(\mathbb{M}^I) = \sum_{n \geq i} \binom{n-1}{i-1} v_n(\mathbb{M}^I) + (-1)^i \lambda(M/IM)$ for every $i \geq 1$.

Proof. It is enough to recall that $g_i(\mathbb{M}^I) = \sum_{j=0}^{i}(-1)^{i-j}e_i(\mathbb{M}^I)$. Now (a) follows by Theorem 2.5 a). Further (b) follows by Theorem 2.5(b) and Northcott's inequality (Theorem 2.4 for $s = 1$) always applied to the filtration $\mathbb{M}^I$. The last part is a consequence of Theorem 2.5(c) and Proposition 5.1. $\square$

Remark 5.1. By Theorem 5.7, it follows that

$$g_1(\mathbb{M}^I) \geq -\lambda(M/IM)$$

We remark that if $g_1(\mathbb{M}^I) = -\lambda(M/IM)$, then $F_I(\mathbb{M})$ does not necessarily have maximal depth. In fact, again by Theorem 5.7 it follows that $gr_{\mathbb{M}^I}(M)$ is Cohen–Macaulay, but nothing is known about $gr_{\mathbb{M}}(M)$. The following example taken from [30] has minimum $g_1(\mathbb{M}^I)$, nevertheless $F_{\mathfrak{m}}(\mathfrak{q})$ is not Cohen–Macaulay.

Consider $A = k[[x^4,x^3y,x^2y^2,xy^3,y^4]]$ as a subring of the formal power series ring $k[[x,y]]$ and let $\mathbb{M}$ be the q-adic filtration with $\mathfrak{q} = (x^4,x^3y,xy^3,y^4)$ and let $I = \mathfrak{m}$. We have $g_1(\mathbb{M}^I) = -1$ since $\mathfrak{m}\mathfrak{q} = \mathfrak{m}J$ where $J = (x^4,y^4)A$. In this case $F_{\mathfrak{m}}(\mathfrak{q})$ has depth 1, hence it is not Cohen–Macaulay.

In the case of the q-adic filtration on M it is possible to characterize the ideals q for which g_1 is minimal. The following result generalizes Proposition 6.1 [42]).

Proposition 5.2. *Let $\mathbb{M}$ be the q-adic filtration on a Cohen–Macaulay module M and let I be an ideal containing $\mathfrak{q}$. Then $\mathbb{M}^I$ has minimal multiplicity if and only if $g_1(\mathbb{M}^I) = -\lambda(M/IM)$.*

Proof. We recall that $\mathbb{M}^I$ has minimal multiplicity if and only if $IM_1 = I\mathfrak{q}M = JIM$ for every ideal J generated by a maximal superficial sequence for $\mathbb{M}^I$. Then by Valabrega–Valla's criterion, $gr_{\mathbb{M}^I}(M)$ is Cohen–Macaulay. Now the result follows by Theorem 5.7(1). since $w_n(\mathbb{M}^I) = 0$ for every $n \geq 1$. Conversely if $g_1(\mathbb{M}^I) = -\lambda(M/IM)$, then in Theorem 5.7(1) we have the equality and $w_n(\mathbb{M}^I) = 0$ for every $n \geq 1$. Hence $gr_{\mathbb{M}^I}(M)$ is Cohen–Macaulay and $IM_n \subseteq J \cap IM_n = JIM_{n-1}$ for every $n \geq 1$. In particular $IM_1 = I\mathfrak{q}M = JIM$, as required. $\square$

Chapter 6
Applications to the Sally Module

W.V. Vasconcelos enlarged the list of blowup algebras by introducing the *Sally module*. Let $(A, \mathfrak{m})$ be a commutative local ring and let $\mathfrak{q}$ be an ideal of A, then the Sally module $S_J(\mathfrak{q})$ of $\mathfrak{q}$ with respect to a minimal reduction J is by definition

$$S_J(\mathfrak{q}) := \oplus_{n \geq 1} \mathfrak{q}^{n+1}/J^n \mathfrak{q}.$$

It is clear that $S_J(\mathfrak{q})$ is a graded module over $\mathscr{R}(J) = \oplus_{n \geq 0} J^n$, the Rees algebra of J. We have an exact sequence of graded $\mathscr{R}(J)$-modules

$$0 \to \mathfrak{q}\mathscr{R}(J) \to \mathfrak{q}\mathscr{R}(\mathfrak{q}) \to S_J(\mathfrak{q}) := \oplus_{n \geq 1} \mathfrak{q}^{n+1}/J^n \mathfrak{q} \to 0.$$

A motivation for its name is the work of Sally where the underlining philosophy is that it is reasonable to expect to recover some properties of $\mathscr{R}(\mathfrak{q})$ (or $gr_{\mathfrak{q}}(A)$) starting from the better structure of $\mathscr{R}(J)$.

Vasconcelos proved that if A is Cohen–Macaulay, then $\dim S_J(\mathfrak{q}) - \dim A$, provided $S_J(\mathfrak{q})$ is not the trivial module.

We extend the definition to modules. As usual $\mathbb{M}$ denotes a good $\mathfrak{q}$-filtration of a module M of dimension r and let J be the ideal generated by a maximal $\mathbb{M}$-superficial sequence for $\mathfrak{q}$. We define

$$S_J(\mathbb{M}) := \oplus_{n \geq 1} M_{n+1}/J^n M_1$$

to be the Sally module of $\mathbb{M}$ with respect to J.

As we saw for the fiber cone, this graded $\mathscr{R}(J)$-module is closely related to the associated graded modules with respect to different filtrations. We consider the J-good filtration induced by the submodule M_1 of M:

$$\mathbb{E} : \{E_0 = M, E_{n+1} = J^n M_1 \quad \forall n \geq 0\}$$

The aim of this chapter is to relate the numerical invariants of $S_J(\mathbb{M})$ to those of the associated graded modules of the filtrations $\mathbb{M}$ and $\mathbb{E}$. As in the previous chapter, we will rediscover easily properties of the Sally module by using the general theory on the associated graded modules developed in the previous chapters.

M.E. Rossi and G. Valla, *Hilbert Functions of Filtered Modules*, Lecture Notes of the Unione Matematica Italiana 9, DOI 10.1007/978-3-642-14240-6_6,
© Springer-Verlag Berlin Heidelberg 2010

6.1 Depth of the Sally Module

The Sally module $S_J(\mathbb{M})$ fits into two exact sequences of graded $\mathscr{R}(J)$-modules with $gr_{\mathbb{M}}(M)$ and $gr_{\mathbb{E}}(M)$.

Proposition 6.1. *Let $\mathbb{M}$ be a good $\mathfrak{q}$-filtration of a module M and let J be the ideal generated by a maximal $\mathbb{M}$-superficial sequence for $\mathfrak{q}$. Then* $\operatorname{depth} gr_{\mathbb{M}}(M) \geq min\{\operatorname{depth} S_J(\mathbb{M}) - 1, \operatorname{depth} gr_{\mathbb{E}}(M)\}$.

Proof. Let $N := \oplus_{n \geq 0} M_n / J^n M_1$, we have the following homogeneous exact sequences of $\mathscr{R}(J)$-graded modules:

$$0 \to gr_{\mathbb{E}}(M) \to N \longrightarrow S_J(\mathbb{M})(-1) \to 0$$

$$0 \to S_J(\mathbb{M}) \to N \longrightarrow gr_{\mathbb{M}}(M) \to 0$$

It is enough to remark that we have the following exact sequences of the homogeneous components of degree n:

$$0 \to J^{n-1} M_1 / J^n M_1 \to M_n / J^n M_1 \to M_n / J^{n-1} M_1 \to 0$$

$$0 \to M_{n+1} / J^n M_1 \to M_n / J^n M_1 \longrightarrow M_n / M_{n+1} \to 0.$$

We remark that $M_n / J^{n-1} M_1 = (S_J(\mathbb{M})(-1))_n$.

The comparison between the depths follow from standard facts (see for example [6]).   □

6.2 The Hilbert Function of the Sally Module

From now on we shall assume that $\lambda(M/\mathfrak{q}M)$ is finite.

If M is Cohen–Macaulay, then the filtration $\mathbb{E}$ is well understood. In fact, since $E_2 = JE_1$, by Theorem 2.9 and Corollary 2.6, $gr_{\mathbb{E}}(M)$ is Cohen–Macaulay with minimal multiplicity and hence

$$P_{\mathbb{E}}(z) = \frac{h_0(\mathbb{M}) + (e_0(\mathbb{M}) - h_0(\mathbb{M}))z}{(1-z)^r}. \tag{6.1}$$

$(h_0(\mathbb{M}) = h_0(\mathbb{E}), e_0(\mathbb{M}) = e_0(\mathbb{E}))$. In particular

$$e_1(\mathbb{E}) = e_0(\mathbb{M}) - h_0(\mathbb{M}) \quad \text{and} \quad e_i(\mathbb{E}) = 0 \text{ for every } i \geq 2. \tag{6.2}$$

Since $\lambda(M_{n+1}/J^n M_1)$ is finite for every n, we may define the Hilbert function of the Sally module

$$H_{S_J(\mathbb{M})}(n) = \lambda(M_{n+1}/J^n M_1)$$

and we denote by $e_i(S_J(\mathbb{M}))$ the corresponding Hilbert coefficients . Starting from the exact sequences of Proposition 6.1, it is easy to get the following equality on the Hilbert series

$$(z-1)P_{S_J(\mathbb{M})}(z) = P_{\mathbb{M}}(z) - P_{\mathbb{E}}(z) \tag{6.3}$$

Several results easily follow from the above equality.

Proposition 6.2. *Let $\mathbb{M}$ be a good $\mathfrak{q}$-filtration of a module M of dimension r and let J be the ideal generated by a maximal $\mathbb{M}$-superficial sequence for $\mathfrak{q}$. We have:*

(1) $\dim S_J(\mathbb{M}) = r$ *if and only if* $e_1(\mathbb{M}) > e_1(\mathbb{E})$.

(2) If $\dim S_J(\mathbb{M}) = r$, *then for every* $i \geq 0$ *we have*

$$e_i(S_J(\mathbb{M})) = e_{i+1}(\mathbb{M}) - e_{i+1}(\mathbb{E}) \tag{6.4}$$

From (6.4) we deduce that the coefficients of the Sally module behave well going modulo $\mathbb{M}$-superficial sequence for $\mathfrak{q}$. We remark that both $\mathbb{M}$ and $\mathbb{E}$ are J-good filtrations, hence by Remark 1.1 it is possible to find in J a sequence of elements which are superficial for both.

The following result was proved in [12] in the particular case of the $\mathfrak{q}$-adic filtration on A. We present here a direct proof in the general setting.

Corollary 6.1. *Let $\mathbb{M}$ be a good $\mathfrak{q}$-filtration of a module M of dimension r and let J be the ideal generated by a maximal $\mathbb{M}$-superficial sequence for $\mathfrak{q}$. Let $\mathbb{N} = \{J^n M\}$ be the J-adic filtration on M and assume $\dim S_J(\mathbb{M}) = r$, then*

$$e_0(S_J(\mathbb{M})) \leq e_1(\mathbb{M}) - e_1(\mathbb{N}) - e_0(\mathbb{M}) + h_0(\mathbb{M}).$$

Proof. It follows from (6.4) and Proposition 2.11.   □

We note that next result rediscovers several results known in the case of the $\mathfrak{q}$-adic filtration on the Cohen–Macaulay ring A (see [113, Corollary 1.2.9, Propositions 1.2.10, 1.3.3]; [83, Corollary 2.7]).

Proposition 6.3. *Let $\mathbb{M}$ be a good $\mathfrak{q}$-filtration of a Cohen–Macaulay module M of dimension r and let J be an ideal generated by a maximal $\mathbb{M}$-superficial sequence for $\mathfrak{q}$, then:*

(1) $\dim S_J(\mathbb{M}) = r$ *if and only if* $e_1(\mathbb{M}) > e_0(\mathbb{M}) - h_0(\mathbb{M})$.

(2) If $\dim S_J(\mathbb{M}) = r$, *then* $e_0(S_J(\mathbb{M})) = e_1(\mathbb{M}) - e_0(\mathbb{M}) + h_0(\mathbb{M})$.

(3) $\operatorname{depth} gr_{\mathbb{M}}(M) \geq \operatorname{depth} S_J(\mathbb{M}) - 1$.

(4) $(z-1)P_{S_J(\mathbb{M})}(z) = P_{\mathbb{M}}(z) - \frac{\lambda(M/M_1) + (e_0(\mathbb{M}) - (\lambda(M/M_1))z}{(1-z)^r}$.

(5) $H_{S_J(\mathbb{M})}(n)$ *is not decreasing.*

Proof. Assertions (1) and (2) follow by Proposition 6.2 and (6.2). Since $gr_{\mathbb{E}}(M)$ is Cohen–Macaulay, then (3) follows by Proposition 6.1 because $\min\{\operatorname{depth} S_J(\mathbb{M})-1,$ $\operatorname{depth} gr_{\mathbb{E}}(M)\} = \operatorname{depth} S_J(\mathbb{M}) - 1$.

The assertion (4) follows from (6.3) and (6.1). Finally (5) follows by (4) and Theorem 1.3.   □

In our general setting we get the following result due, in the classical case, to W. Vasconcelos in [108, Sect. 5.2].

Corollary 6.2. *Let L be a submodule of a Cohen–Macaulay module M of dimension r and let $\mathbb{M} = \mathbb{M}_L$ be a good $\mathfrak{q}$-filtration induced by L. Let J be an ideal generated by a maximal $\mathbb{M}$-superficial sequence for $\mathfrak{q}$, then:*

(1) $\dim S_J(\mathbb{M}) = r$ *provided it is not the trivial module.*

(2) $e_0(S_J(\mathbb{M})) = e_1(\mathbb{M}) - \lambda(L/JM)$ *and* $e_i(S_J(\mathbb{M})) = e_{i+1}(\mathbb{M})$ *for every $i > 0$.*

Proof. By Proposition 6.3, $\dim S_J(\mathbb{M}) = r$ provided $e_1(\mathbb{M}) > e_0(\mathbb{M}) - h_0(\mathbb{M})$. By Theorem 2.9 and Corollary 2.6, this means $M_2 \neq JM_1$. On the other hand, $M_{n+1} = \mathfrak{q}^n L$ for every $n \geq 0$, hence it is easy to see that $S_J(\mathbb{M})$ is the trivial module if and only if $M_2 = JM_1$ and (1) follows. Now (2) is a consequence of Proposition 6.2 since $e_1(\mathbb{E}) = e_0(\mathbb{E}) - h_0(\mathbb{E}) = e_0(\mathbb{M}) - h_0(\mathbb{M}) = \lambda(L/JM)$ and $e_i(\mathbb{E}) = 0$ for $i \geq 2$.
□

Remark 6.1. If $\mathfrak{q}$ is an $\mathfrak{m}$-primary ideal of a local Cohen–Macaulay ring $(A,\mathfrak{m})$ of dimension r, the value of $e_1(\mathfrak{q})$ has a strong influence on the structure of the Sally module. By Corollary 6.2 and Theorem 2.9, if $e_1(\mathfrak{q}) = e_0(\mathfrak{q}) - \lambda(A/\mathfrak{q})$, then $S_J(\mathfrak{q})$ is the trivial module. The case $e_1(\mathfrak{q}) = e_0(\mathfrak{q}) - \lambda(A/\mathfrak{q}) + 1$ is much more difficult. Recently S. Goto, K. Nishida, K. Ozeki proved that, under this assumption, there exists a positive integer $c \leq r$ such that

$$S_J(\mathfrak{q}) \simeq (x_1,\ldots,x_c) \subseteq [\mathscr{R}(J)/\mathfrak{m}\mathscr{R}(J)] \simeq A/\mathfrak{m}[x_1,\ldots,x_r]$$

as graded $\mathscr{R}(J)$-modules. When this is the case $c = v_1(\mathfrak{q}) = \lambda(\mathfrak{q}^2/J\mathfrak{q})$ (see [33, Theorem 1.2]). By using this surprising information and Proposition 6.3(3), we easily obtain

$$P_{\mathfrak{q}}(z) = \frac{\lambda(A/\mathfrak{q}) + (e_0(\mathfrak{q}) - \lambda(A/\mathfrak{q}) - c)z + \sum_{i=2}^{c+1}(-1)^i\binom{c+1}{i}z^i}{(1-z)^r}$$

We remark that, if $r = 2$, we obtain the Hilbert series described in Theorem 4.5. Very recently S. Goto and K. Ozeki announced an extension of the above result relaxing the requirement for the Cohen–Macaulayness of A [34].

Under the assumptions of Corollary 6.2, we remark that if $J = (a_1,\ldots,a_r)$ and $S_J(\mathbb{M})$ is not the trivial module, then the ideal $JT = (a_1T,\ldots,a_rT)$ in the Rees algebra $\mathscr{R}(J) = A[JT]$ is generated by a system of parameters for $S_J(\mathbb{M})$. In fact $S_J(\mathbb{M})/JTS_J(\mathbb{M}) = \oplus_{n\geq 1}M_{n+1}/JM_n$ which is an Artinian module.

In particular $S_J(\mathbb{M})$ is Cohen–Macaulay if and only if $a_1T,\ldots,a_rT$ is a regular sequence on $S_J(\mathbb{M})$.

Theorem 6.1 ([113], Theorem 2.1.6, Corollaries 2.1.7, 2.1.8, 2.1.9). *Let L be a submodule of a Cohen–Macaulay module M of dimension r and let* $\mathbb{M} = \mathbb{M}_L$ *be a good* $\mathfrak{q}$*-filtration induced by L. Denote by J the ideal generated by a maximal* $\mathbb{M}$*-superficial sequence for* $\mathfrak{q}$*, then*

$$e_0(S_J(\mathbb{M})) \leq \sum_{j \geq 1} v_j(\mathbb{M}).$$

The following facts are equivalent:

(1) $e_0(S_J(\mathbb{M})) = \sum_{j \geq 1} v_j(\mathbb{M})$

(2) $e_1(\mathbb{M}) = \sum_{j \geq 0} v_j(\mathbb{M})$

(3) $\operatorname{depth} gr_{\mathbb{M}}(M) \geq r - 1$

(4) $P_{\mathbb{M}}(z) = \frac{\lambda(M/M_1) + \sum_{j \geq 1}(v_{j-1}(\mathbb{M}) - v_j(\mathbb{M}))z^j}{(1-z)^r}$

(5) $P_{S_J(\mathbb{M})}(z) = \frac{\sum_{j \geq 1} v_j(\mathbb{M})z^j}{(1-z)^r}$

(6) $S_J(\mathbb{M})$ *is Cohen–Macaulay.*

Proof. By Corollary 6.2 we have $e_0(S_J(\mathbb{M})) = e_1(\mathbb{M}) - \lambda(L/JM)$. Hence, by Theorem 2.5, we get $e_0(S_J(\mathbb{M})) \leq \sum_{j \geq 0} v_j(\mathbb{M}) - \lambda(L/JM) = \sum_{j \geq 1} v_j(\mathbb{M})$. The equality holds if and only if $e_1(\mathbb{M}) = \sum_{j \geq 0} v_j(\mathbb{M})$. Hence, by Theorem 2.5, the assertions (1), (2), (3), (4) are equivalent. By Proposition 6.3(4), (4) and (5) are equivalent. Since (6) implies (3) by Proposition 6.3(3), it is enough to prove that (5) implies (6).

We may assume $S_J(\mathbb{M})$ has dimension r. We recall that $S_J(\mathbb{M})$ is a $\mathscr{R}(J)$-module and we have $S_J(\mathbb{M})/JTS_J(\mathbb{M}) = \oplus_{n>1} M_{n+1}/JM_n$. From (5) we deduce that $P_{S_J(\mathbb{M})}(z) = \frac{1}{(1-z)^r} P_{S_J(\mathbb{M})/JTS_J(\mathbb{M})}(z)$. Then JT is generated by a regular sequence of length $r = \dim S_J(\mathbb{M})$ and hence $S_J(\mathbb{M})$ is Cohen–Macaulay. $\square$

In the particular case of the $\mathfrak{m}$-adic filtration on A, we can easily give a partial extension of the above result without assuming the Cohen–Macaulayness of A (see [85, Theorem 3.2]).

Theorem 6.2. *Let* $(A, \mathfrak{m})$ *be a local ring of dimension r and let J be an ideal generated by a maximal superficial sequence for* $\mathfrak{m}$*. If* $\dim S_J(\mathfrak{m}) = r$*, then*

$$e_0(S_J(\mathfrak{m})) \leq \sum_{j \geq 0} v_j(\mathfrak{m}) - e_0(\mathfrak{m}) + 1$$

The following facts are equivalent:

(1) $e_0(S_J(\mathfrak{m})) = \sum_{j \geq 0} v_j(\mathfrak{m}) - e_0(\mathfrak{m}) + 1$

(2) $e_1(\mathfrak{m}) - e_1(J) = \sum_{j \geq 0} v_j(\mathfrak{m})$

(3) A *is Cohen–Macaulay and* $\operatorname{depth} gr_{\mathfrak{m}}(A) \geq r - 1$

Proof. It follows by Corollary 6.1 and Theorem 2.6. $\square$

References

1. Abhyankar S., Local rings of high embedding dimension, Am. J. Math. 89 (1967), 1073–1077.
2. Atiyah M.F., Macdonald I.G., Introduction to Commutative Algebra, Addison-Wesley, Reading, MA, 1969.
3. Bertella V., Hilbert function of local Artinian level rings in codimension two, J. Algebra 321(5) (2009), 1429–1442.
4. Bondil R., Geometry of superficial elements, Ann. Fac. Sci. Toulouse, Ser. 6, 14(2) (2005), 185–200.
5. Bondil R., Le' Dung Trang, Characterisations des elements superficials d'un ideal, C. R. Acad. Sci. Paris 332, Ser. I (2001), 717–722.
6. Bruns W., Herzog J., Cohen–Macaulay Rings, revised edition, Cambridge Studies in Advanced Mathematics, Vol. 39, Cambridge University Press, Cambridge (1998).
7. Capani A., Niesi G., Robbiano L., CoCoA, a system for doing Computations in Commutative Algebra, available via anonymous ftp from: cocoa.dima.unige.it.
8. Colomé G., Elias J., Bigraded structures and the depth of blow-up algebras, Proc. R. Soc. Edinburgh Sect. A 136(6) (2006), 1175–1194.
9. Conca A., De Negri E., Jayanthan A.V., Rossi M.E., Graded rings associated with contracted ideals, J. Algebra 284 (2005), 593–626.
10. Conca A., De Negri E., Rossi M.E., Integrally closed and componentwise linear ideals in the polynomial ring, Math. Z. 265(3) (2010), 715–734.
11. Conti D., Successioni superficiali e riduzioni per filtrazioni stabili, Tesi di Laurea, Università degli Studi di Pisa (2006).
12. Corso A., Sally modules of m-primary ideals in local rings, Commun. Algebra 37(12) (2009), 4503–4515, arXiv:math.AC/0309027.
13. Corso A., Polini C., Hilbert coefficients of ideals with a view toward blowing-up algebras, In: I. Peeva (ed.) Syzygies and Hilbert Functions, Chapter 2, Lecture Notes in Pure and Applied Mathematics, Vol. 254, Chapman Hall / CRC, Boca Raton, FL (2007).
14. Corso A., Polini C., Rossi M.E., Depth of associated graded rings via Hilbert coefficients of ideals, J. Pure Appl. Algebra 201 (2005), 126–141.
15. Corso A., Polini C., Vasconcelos W.V., Multiplicity of the special fiber of blowups, Math. Proc. Camb. Phil. Soc. 140 (2006), 207–219.
16. Corso A., Polini C., Vaz Pinto M., Sally modules and associated graded rings, Commun. Algebra 26(8) (1998), 2689–2708.
17. Cortadellas T., Fiber cones with almost maximal depth, Commun. Algebra 33(3) (2005), 953–963.
18. Cortadellas T., Zarzuela S., On the depth of the fiber cone of filtrations, J. Algebra 198(2) (1997), 428–445.
19. Cuong N.T., Schenzel P., Trung N.V., Über verallgemeinerte Cohen–Macaulay-Modulen, Math. Nachr. 85 (1978), 57–73.

20. Elias J., Characterization of the Hilbert–Samuel polynomial of curve singularities, Comp. Math. 74 (1990), 135–155.
21. Elias J., The conjecture of Sally on the Hilbert function for curve singularities, J. Algebra 160 (1993), 42–49.
22. Elias J., On the depth of the tangent cone and the growth of the Hilbert function, Trans. Am. Math. Soc. 351 (1999), 4027–4042.
23. Elias J., On the computation of Ratliff–Rush closure, J. Symbolic Comput. 37(6) (2004) 717–725.
24. Elias, J., Upper bounds of Hilbert coefficients and Hilbert functions, Math. Proc. Camb. Phil. Soc. 145 (2008), 1–8.
25. Elias J., Rossi M.E., Valla G., On the coefficients of the Hilbert polynomial, J. Pure Appl. Algebra 108 (1996), 35–60.
26. Elias J., Valla G., Rigid Hilbert functions, J. Pure Appl. Algebra 71 (1991), 19–41.
27. Fillmore, J.P., On the coefficients of the Hilbert–Samuel polynomial, Math. Z. 97 (1967), 212–228.
28. Ghezzi, L., Goto, S., Hong, J., Ozeki, K., Phuong T.T., Vasconcelos W.V., Cohen–Macaulayness versus the vanishing of the first Hilbert coefficient of parameter ideals (2009), arXiv:1002.0391.
29. Ghezzi, L., Hong, J., Vasconcelos W.V., The signature of the Chern coefficients of local rings, Math. Res. Lett. 16(2) (2009), 279–289.
30. Goto S., Buchsbaumness in Rees algebras associated to ideals of minimal multiplicity, J. Algebra 213(2) (1999), 604–661.
31. Goto S., Nishida K., Hilbert coefficients and Buchsbaumness of associated graded rings, J. Pure Appl. Algebra 181(1) (2003), 61–74.
32. Goto S., Nishida K., Ozeki K., Sally modules of rank one, Michigan Math. J. 57 (2008), 359–381.
33. Goto S., Nishida K., Ozeki K., The structure of Sally modules of rank one, Math. Res. Lett. 15(5) (2008), 881–892.
34. Goto S., Ozeki K., The structure of Sally modules of rank one, non-Cohen–Macaulay cases, J. London Math. Soc. (2009), arXiv:1002.0391.
35. Goto S., Ozeki K., Buchsbaumness in local rings possessing constant first Hilbert coefficients of parameters, Nagoya Math. J. 199 (2010), 95–105.
36. Guerrieri A., On the depth of the associated graded ring of an m-primary ideal, J. Algebra 167 (1994), 745–757.
37. Guerrieri A., On the depth of the associated graded ring, Proc. Am. Math. Soc. 123(1) (1995), 11–20.
38. Guerrieri A., Rossi M.E., Hilbert coefficients of Hilbert filtrations, J. Algebra 199(1) (1998), 40–61.
39. Guerrieri A., Rossi M.E., Estimates on the depth of the associated graded ring, J. Algebra 211 (1999), 457–471.
40. Jayanthan A.V., Puthenpurakal,T.J., Verma J.K., On fiber cones of m-primary ideals, Canad. J. Math. 59(1) (2007), 109–120.
41. Jayanthan A.V., Singh B., Verma J.K., Hilbert coefficients and depth of form rings, Commun. Algebra 32(2004), 1445–1452.
42. Jayanthan A.V., Verma J.K., Hilbert coefficients and depth of fiber cones, J. Pure Appl. Algebra 201(1–3) (2005), 97–115.
43. Jayanthan A.V., Verma J.K., Fiber cones of ideals with almost minimal multiplicity, Nagoya Math. J. 177 (2005), 155–179.
44. Heinzer W., Johnston B., Lantz B.D., Shah K., The Ratliff–Rush ideals in a Noetherian ring, Commun. Algebra 20(2) (1992), 591–622.
45. Heinzer W., Lantz B.D., Shah K., The Ratliff–Rush Ideals in a Noetherian Ring: A Survey, In: Methods in Module Theory (Colorado Springs, CO, 1991), Lecture Notes in Pure and Applied Mathematics, Vol. 140, Dekker, New York (1993), pp. 149–159.
46. Herzog J., Rossi M.E., Valla G., On the depth of the symmetric algebra, Trans. Am. Math. Soc. 296(2) (1986), 577–606.

47. Herzog J., Waldi R., A note on the Hilbert function of a one-dimensional Cohen–Macaulay ring, Manuscripta Math. 16(3) (1975), 251–260.
48. Hong J., Ulrich B., Specialization and integral closure, preprint (2009).
49. Huckaba S., A d-dimensional extension of a lemma of Huneke's and formulas for the Hilbert coefficients, Proc. Am. Math. Soc. 124(5) (1996), 1393–1401.
50. Huckaba S., On associated graded rings having almost maximal depth, Commun. Algebra 26(3) (1998), 967–976.
51. Huckaba S., Huneke C., Normal ideals in regular rings, J. Reine Angew. Math. 510 (1999), 63–82.
52. Huckaba S., Marley T., Depth properties of Rees algebra and associated graded ring, J. Algebra 156 (1993), 259–271.
53. Huckaba S., Marley T., Hilbert coefficients and the depths of associated graded rings, J. London Math. Soc, 56 (1997), 64–76.
54. Huneke C., Hilbert functions and symbolic powers, Michigan Math J. 34(2) (1987), 293–318.
55. Huneke C., Sally J., Birational extensions in dimension two and integrally closed ideals, J. Algebra 115 (1988), 481–500.
56. Huneke C., Swanson I., Integral Closure of Ideals, Rings, and Modules, London Mathematical Lecture Note Series, Vol. 336, Cambridge University Press, Cambridge (2006).
57. Itoh S., Coefficients of normal Hilbert polynomials, J. Algebra 150(1) (1992), 101–117.
58. Itoh S., Hilbert coefficients of integrally closed ideals, J. Algebra 176 (1995), 638–652.
59. Kinoshita Y., Nishida K., Sakata K., Shinya R., An upper bound on the reduction number of an ideal, arXiv:0711.4880v1 [math.AC].
60. Kirby D., A note on superficial elements of an ideal in a local ring, Q. J. Math. Oxford (2), 14 (1963), 21–28.
61. Lipman J., Teissier B., Pseudorational local rings and a theorem of Briancon–Skoda about integral closures of ideals. Michigan Math. J. 28(1) (1981), 97–116.
62. Matlis E., One-dimensional Cohen–Macaulay rings, Lecture Notes in Mathematcs, Springer, Berlin (1973).
63. Narita M., A note on the coefficients of Hilbert characteristic functions in semi-regular local rings, Proc. Cambridge Philos. Soc., 59 (1963), 269–275.
64. Northcott D.G., A note on the coefficients of the abstract Hilbert function, J. London Math. Soc., 35 (1960), 209–214.
65. Northcott D.G., Lessons on Rings, Modules and Multiplicities, Cambridge University Press, Cambridge (1968).
66. Ooishi A., Δ-genera and sectional genera of commutative rings, Hiroshima Math. J. 17 (1987), 361–372.
67. Polini C., A filtration of the Sally module and the associated graded ring of an ideal, Commun. Algebra 28(3) (2000), 1335–1341.
68. Polini C., Ulrich B., Vasconcelos W.V., Normalization of ideals and Briançon–Skoda numbers, Math. Res. Lett. 12 (2005), 827–842.
69. Puthenpurakal T., Hilbert-coefficients of a Cohen–Macaulay module, J. Algebra 264(1) (2003), 82–97.
70. Puthenpurakal T., Ratliff–Rush filtration, regularity and depth of higher associated graded modules, Part I, J. Pure Appl. Algebra, 208 (2007), 159–176.
71. Puthenpurakal T., Ratliff–Rush filtration, regularity and depth of higher associated graded modules, Part II, math.AC/0808.3258v1 (2008).
72. Puthenpurakal T., Zulfeqarr, Ratliff–Rush filtrations associated with ideals and modules over a Noetherian ring, J. Algebra 311(2) (2007), 551–583.
73. Ratliff L.J., Rush D., Two notes on reductions of ideals, Indiana Univ. Math. J. 27 (1978) 929–934.
74. Rees D., α-transforms of local rings and a theorem on multiplicities of ideals, Proc. Camb. Phil. Soc. 57 (1961), 8–17.
75. Rossi M.E., A bound on the reduction number of a primary ideal, Proc. Am. Math. Soc. 128(5) (1999), 1325–1332.

76. Rossi M.E., Primary ideals with good associated graded ring, J. Pure Appl. Algebra 145(1) (2000), 75–90.
77. Rossi M.E., Hilbert Functions of Cohen–Macaulay local rings, School in Commutative Algebra and its Connections to Geometry, Universidade Federal de Pernambuco, Olinda (Brasil) in honor of V.W. Vasconcelos (2009).
78. Rossi M.E., Swanson I., Notes on the behavior of the Ratliff–Rush filtration, Commutative algebra (Grenoble/Lyon, 2001), In: Contemporary Mathematics, Vol. 331, AMS, Providence, RI (2003), pp. 313–328.
79. Rossi M.E., Trung N.V., Valla G., Castelnuovo–Mumford regularity and extended degree, Trans. Am. Math. Soc. 355(5) (2003), 1773–1786.
80. Rossi M.E., Valla G., A conjecture of J. Sally, Commun. Algebra 24(13) (1996), 4249–4261.
81. Rossi M.E., Valla G., Cohen–Macaulay local rings of dimension two and an extended version of a conjecture of J. Sally, J. Pure Appl. Algebra 122(3) (1997), 293–311.
82. Rossi M.E., Valla G., Cohen–Macaulay local rings of embedding dimension $e + d - 3$, Proc. London Math. Soc. 80 (2000), 107–126.
83. Rossi M.E., Valla G., Vasconcelos W.V., Maximal Hilbert Functions, Result. Math. 39 (2001) 99–114.
84. Rossi M.E., Valla G., The Hilbert function of the Ratliff–Rush filtration, J. Pure Appl. Algebra, 201 (2005), 25–41.
85. Rossi M.E., Valla G., On the Chern number of a filtration, Rend. Sem. Mat. Univ. Padova, 121 (2009).
86. Sally J., Bounds for numbers of generators for Cohen–Macaulay ideals, Pacific J. Math. 63 (1976), 517–520.
87. Sally J., On the associated graded ring of a local Cohen–Macaulay ring, J. Math. Kyoto Univ. 17(1) (1977), 19–21.
88. Sally J., Stretched Gorenstein rings, J. London Math. Soc. 20(2) (1979), 19–26.
89. Sally J., Good embedding dimensions for Gorenstein singularities, Math. Ann. 249 (1980), 95–106.
90. Sally J., Reductions, local cohomology and Hilbert functions of local rings, In: Commutative algebra (Durham, 1981), London Math. Soc. Lecture Note Series, Vol. 72, Cambridge University Press, Cambridge (1982), pp. 231–241.
91. Sally J., Cohen–Macaulay local rings of embedding dimension $e + d - 2$, J. Algebra 83(2) (1983), 393–408.
92. Sally J., Hilbert coefficients and reduction number 2, J. Algebraic Geom. 1(2) (1992), 325–333.
93. Sally J., Ideals whose Hilbert function and Hilbert polynomial agree at $n = 1$, J. Algebra 157(2) (1993), 534–547.
94. Sally J., Vasconcelos W.V., Stable rings, J. Pure Applied Algebra 4 (1974), 319–336.
95. Samuel P., Algèbre locale, Mém. Sci. Math. 123 (1953), Paris.
96. Shah K., On the Cohen–Macaulayness of the Fiber Cone of an Ideal, J. Algebra 143 (1991), 156–172.
97. Singh B., Effect of a permissible blowing-up on the local Hilbert functions, Invent. Math. 26 (1974), 201–212.
98. Srinivas V., Trivedi V., On the Hilbert function of a Cohen–Macaulay local ring, J. Algebraic Geom. 6(4) (1997), 733–751.
99. Stückrad J., Vogel W., Buchsbaum Rings and Applications, Springer, Berlin (1986).
100. Trivedi V., Hilbert functions, Castelnuovo–Mumford regularity and uniform Artin–Rees numbers, Manuscripta Math. 94(4) (1997), 485–499.
101. Trung N.V., Reduction exponent and degree bound for the defining equations of graded rings, Proc. Am. Math. Soc. 101(2) (1987), 229–236.
102. Trung N.V., Filter-regular sequences and multiplicity of blow-up rings of ideals of the principal class, J. Math. Kyoto Univ. 33(3) (1993), 665–683.
103. Trung N.V., Verma J.K., Mixed multiplicities of ideals versus mixed volumes of polytopes, Trans. Am. Math. Soc. 359 (2007), 4711–4727.

104. Trung N.V., Verma J.K., Hilbert functions of multigraded algebras, mixed multiplicities of ideals and their applications (2008), arXiv:0802.2329v1 [math.AC].
105. Valabrega P., Valla G., Form rings and regular sequences, Nagoya Math. J. 72 (1978), 93–101.
106. Valla G., On forms rings which are Cohen–Macaulay, J. Algebra 59 (1979), 247–250.
107. Valla G., Problems and results on Hilbert functions of graded algebras. Six lectures on commutative algebra (Bellaterra, 1996), In: Progress in Mathematics, Vol. 166, BirkhSuser, Basel (1998), pp. 293–344.
108. Vasconcelos W.V., Hilbert functions, analytic spread, and Koszul homology, Contemp. Math. 159 (1994), 410–422.
109. Vasconcelos W.V., Arithmetic of Blowup Algebras, London Mathematical Society, Lecture Note Series 195, Cambridge University Press, Cambridge (1994).
110. Vasconcelos W.V., Integral Closure, London Mathematical Society, Springer Monographs in Mathematics, Springer, Berlin (2005).
111. Vasconcelos W.V., The Chern numbers of local rings, Michigan Math. J. 57 (2008), 725–743.
112. Vaz Pinto M., Structure of Sally modules and Hilbert functions, Ph. D. Thesis, Rutgers University (1995).
113. Vaz Pinto M., Hilbert functions and Sally modules, J. Algebra 192 (1997), 504–523.
114. Verma J.K., Hilbert Coefficients and Depth of the Associated Graded Ring of an Ideal, arXiv:0801.4866 [math.AC].
115. Wang H.J., On Cohen–Macaulay local rings with embedding dimension $e + d - 2$, J. Algebra 190(1) (1997), 226–240.
116. Wang H.J., Hilbert coefficients and the associated graded rings, Proc. Am. Math. Soc. 128(4) (2000), 785–801.
117. Zariski O., Samuel P., Commutative Algebra, Vol. II, Van Nostrand, New York (1960).

Index

Editor in Chief: Franco Brezzi

Editorial Policy

1. The UMI Lecture Notes aim to report new developments in all areas of mathematics and
    their applications - quickly, informally and at a high level. Mathematical texts analysing
    new developments in modelling and numerical simulation are also welcome.

2. Manuscripts should be submitted to
    Redazione Lecture Notes U.M.I.
    umi@dm.unibo.it
    and possibly to one of the editors of the Board informing, in this case, the Redazione about
    the submission. In general, manuscripts will be sent out to external referees for evaluation.
    If a decision cannot yet be reached on the basis of the first 2 reports, further referees may be
    contacted. The author will be informed of this. A final decision to publish can be made only
    on the basis of the complete manuscript, however a refereeing process leading to a
    preliminary decision can be based on a pre-final or incomplete manuscript. The strict
    minimum amount of material that will be considered should include a detailed outline
    describing the planned contents of each chapter, a bibliography and several sample chapters.

3. Manuscripts should in general be submitted in English. Final manuscripts should contain
    at least 100 pages of mathematical text and should always include
    - a table of contents;
    - an informative introduction, with adequate motivation and perhaps some historical
      remarks: it should be accessible to a reader not intimately familiar with the topic treated;
    - a subject index: as a rule this is genuinely helpful for the reader.

4. For evaluation purposes, please submit manuscripts in electronic form, preferably as pdf- or
    zipped ps-files. Authors are asked, if their manuscript is accepted for publication, to use the
    LaTeX2e style files available from Springer's web-server at
    ftp://ftp.springer.de/pub/tex/latex/svmonot1/   for monographs
    and at
    ftp://ftp.springer.de/pub/tex/latex/svmultt1/   for multi-authored volumes

5. Authors receive a total of 50 free copies of their volume, but no royalties. They are entitled
    to a discount of 33.3% on the price of Springer books purchased for their personal use, if
    ordering directly from Springer.

6. Commitment to publish is made by letter of intent rather than by signing a formal contract.
    Springer-Verlag secures the copyright for each volume. Authors are free to reuse material
    contained in their LNM volumes in later publications: A brief written (or e-mail) request for
    formal permission is sufficient.